Creative Activities in Mathematics

PROBLEM-BASED MATHS
INVESTIGATIONS FOR
UPPER PRIMARY

BOOK 2

BY DEREK HOLTON AND DUNCAN SYMONS

Published in 2025 by Amba Press, Melbourne, Australia
www.ambapress.com.au

First published in 2015 by ACER Press, an imprint of
Australian Council for Educational Research Ltd

© 2025 Derek Holton and Duncan Symons

All rights reserved. No part of this book may be reproduced or transmitted in any form or by any means, electronic or mechanical, including photocopying, recording or by any information storage and retrieval system, without prior permission in writing from the publisher.

Cover design: Tess McCabe
Editor: Diane Fowler
Typesetting: Peter Long

ISBN: 9781923569249 (pbk)
ISBN: 9781923569256 (ebk)

A catalogue record for this book is available from the National Library of Australia.

TABLE OF CONTENTS

Introduction: The map and the journey   5

Part 1: Number and Algebra   11
    Chapter 1: Hours in a day   12
    Chapter 2: Magimaths   26
    Chapter 3: Mr Mac's iPhone   40

Part 2: Measurement and Geometry   51
    Chapter 4: Hubcaps   52
    Chapter 5: Goats and wheels   67
    Chapter 6: Boxes and more boxes   84

Part 3: Statistics and Probability   97
    Chapter 7: Sports shots   98
    Chapter 8: Roman gamblers   112
    Chapter 9: The school fete   130

INTRODUCTION: THE MAP AND THE JOURNEY

Who is this book for?

This book aims to provide extended activities that either introduce or prepare the way for material from the upper primary section of the Australian Curriculum: Mathematics, or that consolidate material that has already been taught. All of the activities provide valuable opportunities for developing students' abilities within the curriculum's proficiency strands. It is for the use of:

- experienced teachers who work regularly in the area of mathematics
- teachers who may not be as confident in maths as in other areas
- relieving teachers needing additional maths support and class resources.

This is the second book of three in the *Creative Activities in Mathematics* series, which covers most of the school years. Although Book 1 is largely for lower primary students and Book 3 for secondary students, some of the material of those books can be used for upper primary students too.

Where are you going?

The journey is more than the map.

If you live in Melbourne and want to drive with your spouse and children to Hervey Bay in Queensland, it is easy to look up a route on the map and set off for two weeks' holiday in warmer climes. But that's the easy part done.

Now you have to decide how to break up the 24-hour drive. Do you take four days or do it all in one go? Four days is a little long and accommodation is expensive, but one straight push through is potentially dangerous even with two drivers. And no matter how many days you drive, the kids will get restless and the journey could develop into some sort of hell. However, if you start in the early hours of the morning and put sleeping children in the car, you could get to the New South Wales border before they wake up. Then you could go to the Western Plains Zoo, hire bikes for the kids and let them use up some of their pent-up energy.

There are many options, but there will always be something unexpected; you might go along the same roads year after year, but it will never be the same journey. And if someone else is travelling to Hervey Bay, their trip is likely to be different again.

In this book we give you a mathematical map, but we don't expect you to slavishly follow it. You might at the start, but we hope you will change it to fit your teaching goals and your students' interests and needs.

Still, some things are sacrosanct. Don't drive on paddocks—stick to the roads. One important rule for the Melbourne–Hervey Bay trip was to involve the children, and this holds true for the journeys in this book. Talk with your students, not just to them; involve them in discussions and let them tell you their thoughts and reasoning. This should help them enjoy where they are going and better learn and understand the mathematics that lies behind each of these activities.

As a result of this philosophy, you'll see phrases like 'Discuss', 'Let your students', 'Ask' and so on throughout the book. These are a sign that the students' role is to work, think, understand and learn. Your role is to prepare the route, get the petrol and steer as needed, but also introduce the students to the route and become involved in their journey. You should act as a facilitator and scaffolder as students learn how to make their own path. There will be times when you will need to tell a student what to do next, but this should only arise when you have tried everything else you can think of.

In daily life, people constantly come across problems that they never encountered in school. Students need to know how to tackle these open situations at least as much as they need to know subject-specific tools. The activities in this book aim to provide situations where students can learn and practise mathematical skills, but also provide opportunities for them to think, reason and be creative—for them to see that maths is more than learning rules.

Although some of the activities are purely mathematical, many are founded on 'real world' settings that involve areas other than mathematics. For example, in Chapter 2 we look at the solar system with obvious links to science; in Chapter 3 we look at symmetry and involve Indigenous Australian art; and in Chapter 4 we link to art, design and economics as students prepare for and participate in a school fete. These connections encourage students to see mathematics not as an isolated event but as something that has valuable applications in a variety of settings.

Exploring the map

The activities here are 'creative' in that they require students to use their own knowledge and creativity to tackle problems that aren't routine. This doesn't mean that they are creating new things that no-one else has ever discovered, but it will mean that they are creating new ideas for themselves that will increase their motivation, learning, understanding and interest in mathematics and in their ability to tackle new situations in life outside the classroom.

The creative environment means that students will go through part or all of the following sequence at various times in order to get their answers.

- Play
- Guess
- Justify (or not)
- Give up
- Go on

In many of the activities students need to *play* with the problem—to work with it to understand what it is asking—before they do anything else. In the early stages it is worth underlining key words in a problem to make sure they tackle what's being asked, rather than what they *think* is being asked. Some students might also panic at this stage if they can't see how they can solve this problem; encourage them to try some examples and experiment to see what might work. In many activities this involves drawing, making tables and so on in order to get started. Let students know that all of these things, especially experimenting, are a major part of what mathematicians do too.

When it comes time to look for a pattern, a *guess* is necessary—perhaps more than one guess before students are able to see the right pattern. Practising mathematicians call these guesses *conjectures* and, if they are working on a very hard problem, maybe only one in a hundred of their conjectures will be true. (The activities here won't need so many guesses.)

After a guess has been made there are three things that can happen. With any luck, someone will be able to *justify* the guess. In Years 4 to 7 you don't expect or need watertight proofs, but your students should be able to give a reason as to why something works even if they can't fill in all the details. Any justification will need to pass three tests: Does it convince the student who suggests it? Does it convince a friend of theirs? And does it convince a more critical audience? So justifications need to be discussed to make sure everyone agrees that they are convinced.

The second thing that might happen after making a guess is the '*or not*' from the list on page 6. What happens if a conjecture is wrong? Hopefully, a student will find an example that contradicts the guess. If the guess is that five things can happen, and someone finds a sixth thing, that means that the guess was wrong and another is needed. Your students may have to go back and do some more playing in order to get a better guess. But this time they have more data to guess with, so the next guess should be better than the previous one. This iterative process will help them home in on the correct answer.

The third possibility is that students can neither justify nor contradict a conjecture. Some activities can be difficult and not everyone will be able to solve everything right away. Again, let students know that mathematicians have this problem too—for example, Goldbach's Conjecture was posed more than 250 years ago and no-one has been able to prove it so far. In this situation, mathematicians and students alike may have to *give up*. You may have to take the lead and teach them a new piece of mathematics, or scaffold their learning to help them find a solution they have almost discovered. If necessary, reassure them that giving up (for now) can actually be a good problem-solving strategy. If you have worked hard on a problem for a while without success, taking a break can lead to a sudden insight that moves you on.

Even if your students have justified their answer there is often more to do—so don't give up, *go on*. What problems can they think of that are like the problem they have just solved? Mathematicians call these *extensions* or *generalisations*. An extension just changes an element of the original problem. A generalisation tries to extend a problem until a result can be found such that the original problem is a special case of the generalisation. Through this process, mathematicians develop mathematics to cover more and more situations.

As the students go through this process, your role is to make sure that they understand the problem as a whole and know the maths required for a solution. You should also make sure that they are not getting too far off a trail that will lead to an answer. Be prepared to supply scaffolding as the activity progresses, and only provide a direct answer as a last resort. The more work students do themselves, the more they learn and understand, and the better this will prepare them for solving future problems both in and out of school.

How to use this book

This book continues on from *Creative Activities in Mathematics Book 1* and provides material for students in Years 4 to 7. At the Year 4 end of the range there is overlap with *Creative Activities in Mathematics Book 1*, so some of those activities may be useful for students in your class.

Each chapter contains three activities, which may be one of two types. Either we take a problem and extend and develop it, or we present a series of problems that relate to the same theme. Whatever the activity, it is divided into four levels that can enhance the learning of each student in your class. Because of this, a level may contain material from more than one year level of the curriculum. Each activity develops through the levels in a series of 'steps' that develop by moving through the curriculum content descriptions and/or requiring more sophisticated work from students.

Each activity is a series of problems that require students to use both their mathematical content knowledge and their ingenuity. In the case of activities that extend and develop concepts that haven't been used by students in earlier years of their study, it may be necessary to start at Level 1 first to make sure that the bases of the development are understood. Older students should be able to pass through Level 1 more quickly than younger students.

You may also want to consider how you might alter some activities to best suit your class. When you look at an activity you might see that the basic idea is sound but want to simplify it (or extend it) in some way. These activities are resources, not straitjackets; alter them as you think necessary to promote engagement and achievement from your students.

The Australian Curriculum: Mathematics is central to these activities. Each activity includes a table that lists relevant content descriptions. All of the activities here have a strong link to the Problem Solving proficiency strand. Some activities also address multiple content strands, the general capabilities and/or the cross-curriculum priorities of the Australian Curriculum.

Activity layout

Each activity has the following format.

- Initial problem
- Background information
- Big ideas
- Suggested resources (if any)
- Problem aims
- Key concepts
- Possible heuristics/strategies
- Special notes
- Levels and steps

The *Initial problem* is the start of the activity for the students and gives some idea of the topic of the activity. This is followed by *Background information*, which ranges over interesting details about the problem, sometimes with an historical note, and how it might be taught. This section aims to give teachers an overview of the activity and how it develops from the initial problem. It sometimes notes any links with similar problems and related ideas. A table sets out the Australian Curriculum: Mathematics content descriptions for each level of each activity. (Note that only part of some descriptors may be covered by a given activity.) This table also sets out the way the problem develops and makes comments on the mathematics that the levels contain. After the background information we list the *Big ideas* of the activity.

It is self-evident what *Suggested resources*, *Problem aims*, *Key concepts* and *Possible heuristics/strategies* are. *Special notes* are used to provide a key definition or idea that is needed in the activity.

The activity is then broken up into its four levels, which develop the problem based around questions and answers. Changes of directions or extensions of the problem are indicated by 'steps'. Throughout each problem there are questions that teachers might ask of students in the course of developing the activities.

Each level problem is completed by a section called 'Where to from here?', which provides questions teachers could ask students, focusing on the big ideas involved at that level. This section provides new ideas to follow up and enables you and your class to enjoy yourself thinking up new problems as extensions from the work of that level.

Additional resources are available at the series website, http://www.acer.edu.au/cam. These include references and web links to related material, plus activity and summary sheets for students that provide a framework for their responses. These connect to the problems of that level and can be printed/copied for the students' use.

PART 1: NUMBER AND ALGEBRA

Part 1 presents three activities centred on the Number and Algebra strand.

Table 1.1: Number and Algebra activities

Problem	Big ideas
Hours in a day	- Ratio and its use in real situations - Day and year - Rounding of decimals
Magimaths	- Collecting and recording data - Using properties of numbers to continue patterns - Generalising from number properties and results of calculations
Mr Mac's iPhone	- Fractions and decimals, their use and relationship - Rounding of decimals - Use of electronic devices: calculators, spreadsheets

Some reminders before you use these tasks in your classroom:

1. The questions in the text are ones you can ask your students. You are likely to be able to produce similar, more immediately relevant ones for your particular students as you work on these activities with them.
2. We have given suggested links to the Years in the Australian Curriculum: Mathematics for all the Levels in each activity but, given that there will be a spread of ability in your classes, you should take these as a guide only. Take the opportunity to encourage every student to the edge of their comfort zone.
3. To take all students further, sometimes you can omit some of the later steps of a Level in favour of the early steps in the following Level.

CHAPTER 1: HOURS IN A DAY

Initial problem

What are a day and a year on Earth? How long are they and why?

Background information

This activity is based on the concept of ratio, but it has links throughout to aspects of Measurement and Geometry.

At Level 1, where the work is based on days and years in the solar system, there is a strong connection to science. Level 1 is aimed at all students from Year 4 up. You may want to integrate this with elements of your science teaching. There's a great opportunity for your class to collect data online about the solar system and how planets move around in it, and then to use this data for class discussions.

At Level 2 we look at what would happen if the world moved to using a 12-hour day rather than a 24-hour day. Various aspects of Level 2 can be considered by Year 4 students, but the whole level should be in the range of Year 5 students and above.

This is followed in Level 3 by looking at the practicalities of a 10-hour day with respect to school timetables. This level will almost certainly take more than one lesson to complete. Level 3 will begin to extend Year 5 students, but is accessible to Year 6 and 7 students.

In the final Level, students are given the opportunity to choose how many hours they would like to have in a day and to consider how this might affect their parents' working day. We expect all students in Years 6 and 7 to be able to cope with Level 4.

The Level 2 activity here is a 'teacher's choice'—you choose the number of hours used as the basis for a day. We present the activity using a framework of a 12-hour day, but you may want to split the day into 8 hours, 6 hours, 4 hours or even 2 hours.

The choice depends on your students' ability. Twelve hours will mean they can handle problems involving division by 2; 8 hours, division by 3; 6 hours, division by 4; 4 hours, division by 6; and 2 hours, division by 12. You could even assign different day-lengths to different students, depending on their ability.

Table 1.2: Australian Curriculum content descriptions for the *Hours in a day* activity

Activity level	Problem	Content descriptions
1	A day in the solar system	*Year 4* Count by quarters, halves and thirds, including with mixed numerals. Locate and represent these fractions on a number line (ACMNA078) Use am and pm notation and solve simple time problems (ACMMG086) *Year 5* Use estimation and rounding to check the reasonableness of answers to calculations (ACMNA099) Solve problems involving division by a one-digit number, including those that result in a remainder (ACMNA101) *Year 6* Add and subtract decimals, with and without digital technologies, and use estimation and rounding to check the reasonableness of answers (ACMNA128)
2	The 12-hour day	*Year 5* ACMNA101 (see above) Compare 12- and 24-hour time systems and convert between them (ACMMG110) *Year 6* Select and apply efficient mental and written strategies and appropriate digital technologies to solve problems involving all four operations with whole numbers (ACMNA123)
3	A 10-hour day	*Year 5* ACMNA101 (see above) *Year 6* ACMNA123 (see above) Interpret and use timetables (ACMMG139)
4	A work YRday	*Year 6* ACMNA123 (see above) ACMMG139 (see above) *Year 7* Express one quantity as a fraction of another, with and without the use of digital technologies (ACMNA155) Round decimals to a specified number of decimal places (ACMNA156) Recognise and solve problems involving simple ratios (ACMNA173)

Big ideas

» How the solar system works, especially the notions of day and year
» Ratio and its use in real situations
» Rounding of decimals, ratio and their use in real situations
» Place value

Problem aims

» Ratio

Key concepts

» Rounding

Possible heuristics/strategies

» Make a table
» Draw a diagram
» Use a calculator
» Go online
» Use digital technology

Special notes

Leap year: A *leap year* is a calendar year when an extra day, 29 February, is inserted to allow for the time it takes the Earth to go round the Sun. *Leaplings* are people who are born on 29 February.

Decimalisation: *Decimalisation* means using a base-10 number system, such as our standard decimal number system. The vast majority of modern cultures use decimal systems for length, volume, weight, temperature and so on, although the United States and a small number of other countries still use some non-decimal measurement systems: inches and feet for length, ounces and pounds for weight, degrees Fahrenheit and so on. Time has not been decimalised in any country; the Level 3 activity tries to give an idea of what such a system would look like.

Level 1: A day in the solar system

Problem

What are a day and a year on Earth? How long are they and why?

Problem steps

Step 1

Hold a discussion about this. We'll talk about a day in Step 4. In the meantime, what is a year is actually a surprisingly difficult question. Some possible responses are 365 days or 366 in a leap year. Strictly speaking a day is the time it takes for the Earth to go once round the sun. But why do we need a leap year? Because a regular year is not equal to exactly 365 days. Due to the vagaries of the solar system, it takes about $365\frac{1}{4}$ days for the Earth to go round the Sun. You might get your class to look this up online.

(To be completely accurate, it actually takes 365.2429 days for the Earth to do a full orbit around the Sun. Even this is not cut and dried, but let's stay with that for now.)

Step 2

Ask the students if anyone is a 'leapling'—that is, someone who was born in a leap year on 29 February. What is a leapling's birthday? How often does a leapling have a birthday? Does that mean that they're only a quarter of the age of everyone else? (See the series website for some useful links.)

Why does 29 February occur only every four years? What do we call these special years that have 29 days in February?

Can the class give some examples of leap years? (2012, 2016, 2020, 2024 etc.) From this pattern, can they see what the rule for a leap year is?

Ask: why do we need to have leap years? Why must these years have a factor of 4? (It was an arbitrary decision, but it does make things a little neater than if 2013, 2017, 2012 etc. were leap years.)

It is that awkward $\frac{1}{4}$ of a year that is the problem—it means there is a discrepancy between the calendar year and the solar year. If we did not have leap years, the year would slowly change so that eventually we would have summer in the 'winter' months. Among other things, this change of seasons would make it difficult for farmers to know when to plant seeds and sow crops.

So we tolerate an extra $\frac{1}{4}$ day where the timed year first falls $\frac{1}{4}$ day short of the actual solar year. In the next year the timed year falls behind by an extra $\frac{1}{4} + \frac{1}{4} = \frac{1}{2}$ day. In the third year of the cycle, the timed year gets even further behind—by $\frac{1}{4} + \frac{1}{4} + \frac{1}{4} = \frac{3}{4}$ days. But in the fourth year we're behind in our time by $\frac{1}{4} + \frac{1}{4} + \frac{1}{4} + \frac{1}{4} = 1$ full day, so we bring it all back to where it started by adding an extra day to the year (see Figure 1.1).

Figure 1.1: The leap year principle

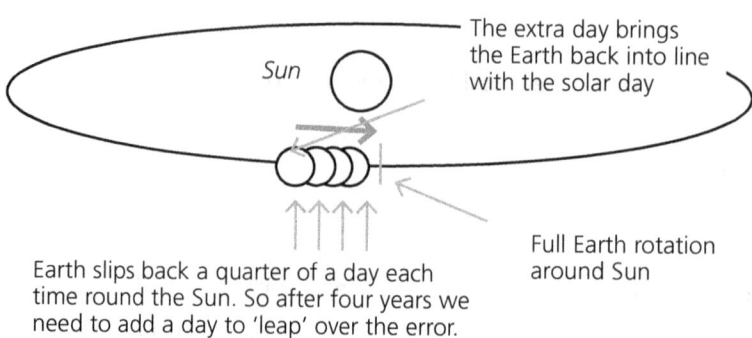

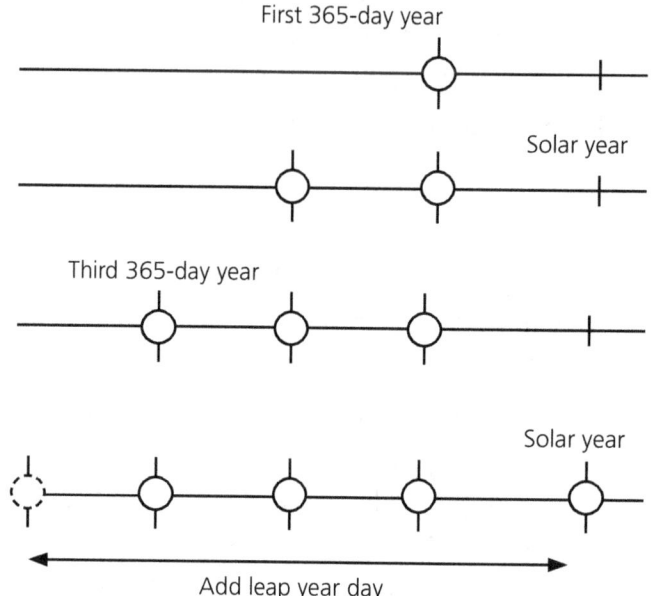

You may want to leave it there, but to extend it further, point out that the '4-year rule' isn't quite right. It would be correct if the Earth took *exactly* $365\frac{1}{4}$ days to complete a solar year, but the solar year is closer to 365.24 days than 365.25. That extra difference has to be accounted for from time to time, so every 100 years we miss out on a leap year so that the Earth can catch up. This means that 2100, 2200, 2300 and so on will not be leap years. But even that is not quite right—to make things even more complicated, any year that is divisible by 400 (such as 2000 and 2400) is *still* a leap year!

Step 3

Tell your students that in a new science fiction movie, an asteroid has hit the Earth and changed its year to *exactly* $365\frac{3}{4}$ days. Which would be the most convenient way to adjust the calendar? Put the students into groups to find at least two ways to solve the leap year problem.

The simplest way is to add three days every four years, so that leap years always have 368 days. An alternative could be to add one day to every year that is divisible by 2 and another day to those years divisible by 4.

Step 4

Now ask your students to think about days. How long is a day on Earth?

This is also more difficult than you might have thought. What do we mean by 'a day'? Do we mean a whole day or the period of daylight? The latter changes continuously over the year as the time of the Sun's rising and setting changes. (You might like to investigate this by keeping track of the Sun's movements during a given term.)

Clarify that you are talking about the whole day, which is 24 hours long. But what is a day? Give your students the chance to talk about how the 24-hour day is defined. It is actually the time it takes for the Earth to turn once on its axis. And even that is not *exactly* 24 hours, so from time to time we have to add in a small amount of time—a *leap second*—to make sure that everything stays in time. We wouldn't want school to start in the middle of the night—or would we?

This is the same concept and difficulty as the leap year problem, but much smaller.

Step 5

Now it's data collection time. Get students to search online for information about the days and years of the other planets in the solar system, and use it to complete the following table. One site that might be useful is Universe Today (see the series website for a link).

Table 1.3: Days and years for the planets of the solar system

Planet	Mercury	Venus	Earth	Mars	Jupiter	Saturn	Uranus	Neptune
Day (in hours)			24					
Year (in Earth years)			1					

What interesting facts have they found? Let the students explain what information interested them most. They might think that the planets with the longest and shortest days and years are interesting; the fact that some planets have days that are longer than their years; or that on some planets the Sun rises in the west and sets in the east.

Follow up the discussion with questions. Which planet goes round the Sun the quickest? Which the slowest? Which planet spins on its axis the fastest? Which the slowest?

Now move the questions about data into mathematical calculations. For instance, which planet moves the fastest in terms of time to complete its orbit? How much faster is it than the others? To answer this, the students will have to find the ratios of the fastest planet to the others.

Chapter 1: Hours in a day

Step 6

One website (see series website for link) gives Jupiter's day as being 9.92496 hours long. Discuss what time would be midday there in hours and minutes. The decimal can be changed to minutes by multiplying by 60. This gives a day of 9 hours and 58 minutes (rounding up), so midday is at 4.29.

This might be a good time to talk about rounding if you have not already done so. Students will need to use rounding again at later levels.

On a sunny day on Jupiter, when daylight is half the time of Jupiter's day and the Sun is at its highest at midday, when does the Sun rise and set? Sunrise will be 2.14.5 and sunset at 6.43.5. (This will happen twice a year at the *equinoxes*, the points when a day has an equal length of day/sunshine and night.)

Step 7

The students might be interested to find out how old they are on other planets in terms of that planet's years. For instance, Mercury's year is about 0.24 of an Earth year, so a 10-year-old from Earth who landed on Mercury would be 10 ÷ 0.24 = 41.67 years old. On the other hand, if you want to get younger you should go to Neptune—although maybe the students don't want to be that young!

Where to from here?

- Let the students choose any day and year lengths they like. Get them to determine a leap year system for that planet and also determine the times of sunrise and sunset on an equinox. It might be interesting to find how often the equinoxes occur.

- If Venus and Earth were in a straight line with the Sun today, when would they next be in a straight line with the Sun?

- How does the Hebrew calendar, account for leap years? (It inserts an additional month, Adar Aleph, seven times in a 19-year cycle.) What about other calendars? What other calendars are there?

- What other problems can your students think up and solve about one or more of the planets?

Level 2: The 12-hour day (teacher's choice)

Problem

Is a day being 24 hours built into the universe, or was the number chosen by someone at some stage in history?

As noted on p. 13, the work at this level can be done just as well by using 8 hours, 6 hours, 4 hours or 2 hours. Your decision depends on how your class can cope with halves, thirds, quarters, sixths or twelfths. If you don't use a 12-hour day, you will need to make a few minor changes to the content of this level.

Problem steps

Step 1

Use this question to promote a class discussion.

Who first decided that 24 hours was a good time for a day? Ask students to try to find this information online. (See the series website for a link to the ABC Science article 'Why are there 24 hours in a day?', a good source of information, but there are others too.) Let your students read these articles for themselves and then report back on what they've found.

Step 2

Suppose Australia decided that the Earth day should no longer be 24 hours. Instead it should be 12 hours long, but each hour will still be divided into 60 minutes. Note that the new minutes will *not* be the same length as the ones we have now. Why not?

Call the 12-hour day system *OZtime*, which is measured in *OZhours* and *OZminutes*. What time would school start in the morning under this new arrangement? Let students work on the problem in groups and then report back for a whole class discussion, explaining what method they used to get the answer. (It should be a matter of dividing the current time by two.)

Step 3

Get students back into their groups for a new OZtime question—how would speed signs need to be changed?

The current signs have numbers such as 40, 50, 60, 70 and 100 on them, signifying a maximum speed in km/hour. In OZhours these would be 40 × 2 = 80; 50 × 2 = 100; 60 × 2 = 120; 70 × 2 = 140; and 100 × 2 = 200. This is because if a car can travel 40 km in one hour of 24-hour time, it will travel 40 km in half-an-OZhour and 80 km in one OZhour.

Have the groups report back and explain why they decided on the numbers they did. They might use some diagrammatic way to help explain.

Why do we divide time by two to get OZtime but multiply distances by two?

Step 4

There are several common ways to use language that involve time—'that shop is open 24/7', 'I did that 20 minutes ago' and so on. Have the class make a list of these expressions, then ask them if they would need any changes under the new OZtime.

For example, 24/7 would become 12/7. Twenty minutes is one-third of an hour, and one OZhour is two old hours. So 20 minutes in old time would be $\frac{1}{3} \times \frac{1}{2} = \frac{1}{6}$ in OZtime; and $\frac{1}{6}$ of the new hour is 10 OZminutes, so the statement becomes 'I did that 10 minutes ago'.

Step 5

The committee assigned to convert Australia to OZtime has decided that doing these calculations all the time will be too hard for most people, so they will produce a conversion table to make it easy for everyone. Your class has been given the contract for creating that table. Divide the contract up between your students by putting them into six groups and getting each to complete one part of Table 1.4.

Table 1.4: The link between OZtime and usual time

24-hour time	0030	0100	0130	0200	0230	0300	0330	0400
OZtime								

24-hour time	0430	0500	0530	0600	0630	0700	0730	0800
OZtime								

24-hour time	0830	0900	0930	1000	1030	1100	1130	1200
OZtime								

24-hour time	1230	1300	1330	1400	1430	1500	1530	1600
OZtime								

24-hour time	1630	1700	1730	1800	1830	1900	1930	2000
OZtime								

24-hour time	2030	2100	2130	2200	2230	2300	2330	2400
OZtime								

Put the completed tables where they can be seen by all students. Get the class to check for consistency—the entries in each successive entry of Table 1.4 should increase by 0.25, or 15 minutes of an OZhour.

Discuss with the whole class what methods they used to fill in the table, and in particular what shortcuts (if any) they took. How many ways can they think of to do the conversion?

Step 6

Suppose that your school ends at 3.30 pm under 24-hour time. What time would this be in OZtime?

This has a few problems. First of all, 3.30 pm is 1530 in 24-hour time, and 0.30 means 30 minutes rather than 0.30 of an hour. So the time 1530 is actually $15\frac{1}{2}$ hours. The OZtime for $15\frac{1}{2}$ hours is $15\frac{1}{2} \div 2 = 7.75$, but the 7.75 is in *decimals* rather than hours and minutes, meaning it is 7 OZhours and 45 OZminutes. Students may find it easier to consider this using fractions, where $7.75 = 7\frac{3}{4}$.

What if your school finishes at 3.15 pm? This looks like being $7\frac{5}{8}$ or 7.625. How is the school going to get someone to ring the bell at exactly 7 OZhours 37.5 OZminutes?

Your students could be faced with many situations like this when OZtimes involve decimals. Do they want to round these to whole minutes? If so, how will they do this approximation? (There is no right answer here, but it's useful to see how students consider and approach the question.)

Discuss how accurate they will need to be in this conversion of the school timetable to OZtime and how they will round all the other times that may turn up.

Step 7

Go back to Step 5 and ask the class if there is a better way to show this time data. What ideas do they have? What are the advantages and disadvantages of the methods they suggest? Would they still have the same advantages if we were converting to 6 hours a day for OZtime?

Step 8

We need a good way to break the OZday into units of time. What do students suggest? How long would be a good time for a working day, for example? What about for the length of a TV program?

Do students think that people will bother to do these calculations after they have used OZtime for a while, or will it just be automatic?

Step 9

Let the class discuss whether or not this move to 12 hours in a day was a good or bad decision. Why, or why not?

Where to from here?

- There are many ways that we use time in our society. Let the class make a list of these that they think could be changed to help people out. How should they be changed?
- Get students to design an analogue watch for OZtime or a speedometer to help people transfer to OZtime in the car.
- In cricket, what does 14.3 overs mean? (Fourteen overs and three balls.) Discuss other places where we use decimal notation for numbers that are not really decimals.

Level 3: A 10-hour day

Problem

In a spirit of decimalisation, the UN has decided that the world should shift to *UNtime*—a 10-hour day with each hour being 100 minutes long.

How can students convert the schedule for a day's programs for a TV channel of their choice to UNtime?

Problem steps

Step 1

You might like to discuss what *decimalisation* means before you start this Level.

If your students have completed the Level 2 activity, it is reasonable to let them take on the task without too much discussion. However, you might like to talk about how specific times will be converted, given that 100 minutes to the hour is an extra complication.

If your class hasn't yet tackled Level 2, you can still follow the outline below, but bear in mind that students have not had a chance to look at similar problems and may require additional scaffolding.

Step 2

Let the students try some simple conversions from 24-hour time to UNtime, first of hours only (e.g. 2 hours, 5 hours, 8 hours) and then of hours and half hours.

The basic principle here is to use the fraction $\frac{10}{24}$ to convert normal hours to UNhours. So 6 hours = $6 \times \frac{10}{24}$ = 2.5 UNhours. But there are problems at the minutes level. If we want to change minutes to UNminutes, first note that 2.4 normal hours = 1 UNhour. Thus, 2.4×60 minutes = 144 minutes = 100 UNminutes; and 1 minute = $\frac{100}{144}$ UNminutes. The ratio we need to change minutes to UNminutes is $\frac{100}{144}$. So 40 minutes = $40 \times \frac{100}{144}$ UNminutes = 27.78 UNminutes.

Step 3

Now let the class think about methods for changing *hours* in the 24-hour system to time in UNtime. They might be guided by the Step 2 conversions. Some students convert normal and UN time to minutes (or hours) and use ratios to change between these.

Step 4

Repeat Step 3, but this time look at changing *minutes* over to UNtime.

Step 5

What rounding would your students do for UNtime so that events don't start or end at awkward times? For example, you may want the school day to start at a whole number of minutes. It might even be a good idea to round the time the bell is rung to the nearest whole five minutes. We leave the level of accuracy to your students.

Step 6

Let the class check that they have a correct conversion method and a correct rounding procedure before continuing to the next step.

Step 7

Get students into groups to address the TV program task.

Each group selects a TV channel and finds the schedule for one day's programming. Using their method and rounding procedure, each group converts the channel's programming to UNtime.

Step 8

Once people have lived with the system for a while they probably won't worry about changing from 24-hour time to UNtime. They'll be more concerned that their TV programs start and end at convenient times rather than at awkward fractions or decimals of times.

Discuss with the class how this can be done. Will programs have to be longer or shorter? Will there need to be more advertisements? What advantages or disadvantages will there be for the TV channel? What advantages or disadvantages will there be for the viewer?

Step 9

As a class, discuss what advantages and disadvantages the change to UNtime would have.

The arithmetic value of 10-hour time is clear; consider trying to add 8 hours 15 minutes and 5 seconds to 17 hours 51 minutes and 58 seconds. But what are the drawbacks?

Who has tried to introduce this system in the past? Why has it not succeeded? Ask students to look online for information or articles about this. They should learn that the French introduced it after the French Revolution; it proved unpopular and was withdrawn a few years later. The Chinese had used it well before that, and it is still used for some purposes by astronomers.

Where to from here?

- What decisions did the students need to make to complete their TV program schedule?
- Where did they have to use ratios? Where did they use rounding? How did they use rounding?

Level 4: A work YRday

Problem

Ask your students to invent a new hourly system for the day. It could be 100 hours long, 48 hours long, 60 hours long or whatever they decide. We'll call this *YRtime*.

Problem steps

Step 1

Get students into groups, each of which creates its own YRtime system. (Alternatively, you could do this as a whole-class activity, with one YRtime system being created and used, but the notes here assume a group activity.)

How does 24-hour time convert to YRtime? Let each group decide on a conversion method, using similar skills and approaches to those used in Levels 2 and 3.

Step 2

The working situations of adults will change as a result of YRtime. Ask each group to choose a particular profession and gather data about the typical working lives and schedules of people in that profession.

Groups then convert the data over to their new YRtime system, drawing up a detailed plan for the working day. How long is the day? What time do people start and end work? When do they have lunch, and for how long? Do they get morning or afternoon breaks; if so, when do they occur and for how long?

Step 3

As a result of an hour being different, pay rates will need to change too. Ask each group to research the average hourly pay rate of those in their chosen profession and convert this rate to the new system.

Workers and their employers will have different ideas about how rounding should be used. What advantages or disadvantages will rounding have for each of them? What dollar difference will this mean over a full year? How can this be avoided? (For example: round down and give holiday bonuses? Round up and get less in superannuation?)

Step 4

Other things will change under the YRtime system. For example, how long would it take to fly from Sydney to London using the new system? Groups need to look online, find out what the current flight time is and use their conversion method to determine the new time.

Step 5

How would your students change things to make daylight saving work? Start by discussing how the current daylight saving system works. (You may need to give additional context if you are in a state that does not have daylight saving.)

Does the current system *have* to use one hour? What problems would arise if the change in time was 2 hours or 30 minutes in the 24-hour system?

What other changes will need to be made for YRtime?

Step 6

Ask each group to design the speedometer for a car to cope with YRtime.

Students need to consider that a car travelling at 100 kph will cover 100 kilometres in one standard hour. If we think about this in UNtime, where one such hour is $\frac{10}{24}$ UNhours, the car will travel 100 kilometres in $\frac{10}{24}$ UNhours. This means that it travels $(100 \times 24) \div 10 = 240$ kilometres in one UNhour, so its speed is 240 kpUNh. Once students understand this, they should be able to convert speeds to YRtime and then design the speedometer.

How will YRtime affect speed restrictions? Get the groups to design new street signs for these new speed restrictions. How will rounding affect life here? (It could mean higher or lower fines will be demanded, depending on the rounding used. What problems will this produce?)

Step 7

Bring the groups back together as a class, and ask the whole class to design a speedometer for a car that can simultaneously cope with *three* of the YRtimes that they have invented.

Where to from here?

- Ask students why they chose the particular YRtime they did. What advantages does it have over 24 hours?
- Ask students for an example of how they used rounding decimals in people's working life. What value is there in rounding in these cases? How did they use ratios to determine the hourly pay of workers?
- What other things in life would need to be changed if the number of hours in a day was changed? What effect would rounding have as a result?

CHAPTER 2: MAGIMATHS

Initial problem

Can someone use mathematics to read your mind?

Background information

This activity looks at situations that appear to let the teacher read the minds of the students. The main question in each of the four levels is whether mind reading is actually involved or if something else is going on. If there is, can we discover what the trick is?

Note: These activities are based on a variety of videos and animations that can be freely found online. We present them with point-by-point instructions for doing them yourself, rather than just using the internet, for several reasons. First, sometimes websites close down, are blocked by firewalls or can't be accessed for some reason. Second, many of these sites use software such as Flash or Java that may not run on every computer (especially on iPads and tablets). And third, these sites may have advertising or other material that you don't want students to see or read. A set of links for these activities can be found at the series website; have a good look at the sites before you use them to make sure that they're appropriate for use.

This activity focuses on magic tricks. One of the points of the activity is to suggest to students that all magic is based on tricks, and many things in the world that are portrayed as 'magic' or quirks of nature actually occur by human intervention. Students may be able to work out some of the easier tricks themselves.

Another point of this activity is to provide some situations that are purely Understanding and Fluency proficiency strand work. The Level 1 and 2 tricks provide this kind of activity, and allow students to play with numeracy in ways that don't strictly connect to the content descriptions of the Australian Curriculum: Mathematics. The Level 3 and Level 4 activities have stronger curriculum links and elements that directly relate to the Reasoning proficiency strand.

The Level 1 trick involves simple sleight-of-hand with cards. Students need to look carefully at what is happening to understand how it works. Collecting data relevant to the trick is important here, as it is in all four levels; before a decision can be made it's always good to know what is happening. Level 1 is available to all students.

Level 2 mixes number tricks with several rearrangements of cards into a given format. Level 2 is also accessible to all students, although it is harder than the trick in Level 1.

Level 3 gets away from cards; the essence of the trick is to turn a number into a multiple of nine in ways that students don't immediately perceive. The trick relies on the test for divisibility by nine. Those students who have been introduced to it will be able to tackle this piece of magic.

At the last Level a process involving arithmetic operations changes any number into a fixed number. This can most easily be done using algebra, but we expect that students in Year 6 (and more able students in Year 5) will be able to work this trick out.

Table 1.5: Australian Curriculum content descriptions for the *Magimaths* activity

Activity level	Problem	Content descriptions
1	The disappearing card	None
2	Guess your number	None
3	Strange signs	*Year 4* Explore and describe number patterns resulting from performing multiplication (ACMNA081)
4	Think of a number	*Year 4* ACMNA081 (see above) *Year 6* Select and apply efficient mental and written strategies and appropriate digital technologies to solve problems involving all four operations with whole numbers (ACMNA123)

Throughout these activities it is important for students to write down what they see so that they can refer to it later; this will help them find the patterns that give the trick away. Encourage students to think about what data they could collect in each problem and which pieces of information might help them. Also encourage students to make up their own tricks, based on what they have seen, to try on their family and friends.

You might like to choose one or two additional mathematical games to slip into the odd lesson, especially if you can match the game with the curriculum material that you are doing at the time.

Big ideas
- Making guesses/conjectures
- Justifying results
- Collecting data

Suggested resources
- Packs of cards
- Computer and large screen/whiteboard display

Problem aims
- To understand some magic tricks
- To use mathematics in real world situations
- To justify mathematical arguments
- To lay a basis for the introduction of algebra

Key concepts
- Conjectures
- Justification
- Collecting data
- Making tables
- Simple subtraction
- Sorting sets of objects
- Division by 9

Possible heuristics/strategies
- Experiment
- Keep records
- Make tables
- Draw diagrams

Special notes

Conjecture: A conjecture or guess is a possible explanation based on incomplete information about a problem; for instance, trying to find a pattern of numbers or objects, or looking for a way that something behaves. In this activity students need to collect data in order to find the common element that always makes the trick work, and use that data to predict the trick's outcome.

Justification: Justification and 'proof of a conjecture' are used almost interchangeably in this book. However, the concept of proof is a little stronger in that it needs to be accepted by an 'enemy', or at least a mathematician. Generally, a justification that shows that a student has grasped the basic ideas behind a formal proof should be considered acceptable. In the *Magimaths* activity, accept a justification that enables a student to solve the problem and develop a similar trick based on the same idea.

Level 1: The disappearing card

Problem

Perform a magic trick for the class, following these steps.

1. Prepare a set of six court or face cards as shown in Figure 1.2, either using physical cards or an image on a computer.
2. Prepare a second set with five court or face cards, also as shown in Figure 1.2. Make sure that *all* of these cards are different to the first six cards.
3. Show the class the six cards.
4. Let a student choose one.
5. Tell them you are going to remove the card that has been chosen.
6. Use patter to distract students' attention away from the set of six cards.
7. Show the set of five cards.

Figure 1.2: Two example sets of cards for the first trick

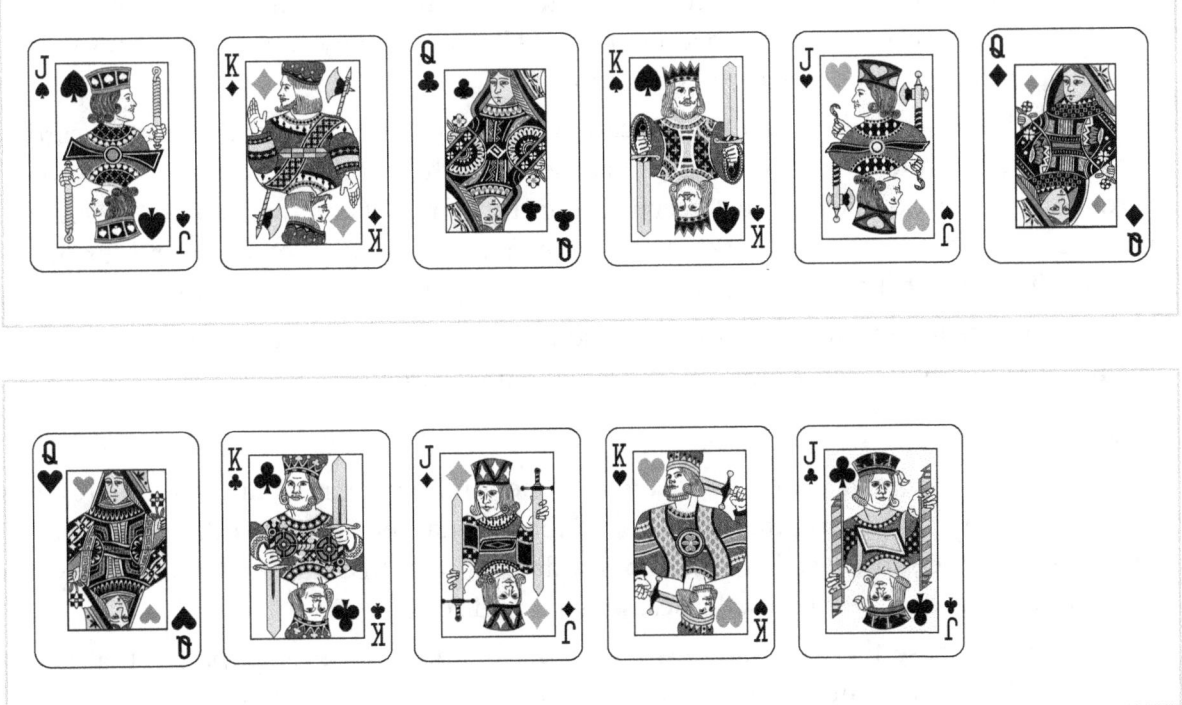

The trick here is that you have removed *all* of the original cards, so you don't need to know what the chosen card was.

If you want to do this trick again, you will have to prepare several different sets of six and five cards.

The easiest way to do this trick yourself is with playing card images that you switch between on a computer, such as by making two slides in a Microsoft PowerPoint display.

Doing it with real cards is harder unless you have good sleight-of-hand skills. One possible option is to arrange the set of five cards, cover it with a sheet of paper, then lay out the set of six cards on top. Once the student has chosen their card, put a screen (such as a book or sheet of paper) down in front of the cards to block students' vision. Remove the set of six cards and take away the paper, leaving the set of five cards showing, then remove the screen.

Problem steps

Step 1

Perform the trick for your students as explained on page 29, either physically or via the link on the series website.

It is useful to get the students to write down the card they have chosen so that they don't forget it.

Can the computer/teacher really read your mind or is there a trick to this? Repeat the trick with another student.

Step 2

Get two students to choose a different card each. How can the computer know what *both* cards are? Get the class to guess what the trick is. Collect as many conjectures as possible. Ask them how they can prove their conjectures.

The cards that appear in the set of five cards are totally different from the original cards that appeared. While the card the student chose is not there, neither are any of the other cards. And because of the delay between looking at the first set and the second set—either because of the way the website works, or the patter and distractions you used if performing it yourself—students have likely forgotten what the original set of cards were.

Get students to see they need to collect some data. Ask: what data? If necessary, ask them what the original cards were. What were the final cards?

Step 3

Get your students to make two sets of cards: one set to show their 'victim' at the start and one set, with one fewer card, that differs from the original set. The cards shouldn't be too different—use the same number values but different suits. Instead of playing cards you might also make cards with dots, dashes, symbols and so on.

Let them practise the trick in class. It is good for them to develop 'patter' that lasts long enough for all of the original cards to be forgotten. Then ask them to try the trick on their parents and friends. They should report back how well it went in the next class. (They might even try the trick on other classes to see if the other class can discover how it works.)

Step 4

For this trick you will need a pack of cards; be sure to practise the trick before you use it in class.

Tell your students that you are going to read their minds. Show your students the top card (which we'll call X) and ask them to remember and record it. Don't look at the card yourself. Now say that you'll read their minds to find the card that they have chosen.

Get a student to choose a small number, such as eight, and put that number of cards face down on a desk, one at a time on top of each other. Then put the cards back on the top of the deck. (In this example, X is now the 8th card in the pack.)

Ask a different student to choose a number, such as 11, that is bigger than the last number (8), but not bigger than the number of cards in the pack. Deal out that number of cards face down on top of each other as before, then put the dealt cards back on the undealt rest of the pack. (At this point X is the 3rd card in the pack.)

Now deal out the difference between the bigger and smaller number (here 11 − 8 = 3). Tell them the card they were thinking about is now on the top of the pack. When you turn over the next card on the pack, it should be card X.

Can anyone see how this trick worked? Let them discuss what is going on. Ask for conjectures and proofs. What data do they need to collect? If anyone says they know how to do the trick, get them to perform it in front of the class.

When most students understand what is happening, get them to perform it in pairs. Ask them to try it on their parents and report back how it went.

Where to from here?

- There are lots more tricks on 'Computer Science 4 Fun', which can be accessed through the series website. You might like to choose one or two to play with the class over a few weeks.
- Some students may also know some card tricks that they could show the rest of the class.
- Ask the students:
  - What conjectures did you make?
  - Which conjectures were false and why?
  - Which conjectures were true? How did you justify them?
  - Were you stuck at any point? How did you get past that difficulty?

Level 2: Guess your number

Problem

Perform a magic trick for the class, following these steps.
1. Download the PDF teacher file for this activity from the series website.
2. Use a projector or interactive whiteboard to show the first page of the PDF (the square of numbers) to the class.
3. Ask one student or a group of students to secretly choose a number.
4. Get one student to tell you the colour that the number has.
5. Use patter to distract the students, stating that the number isn't coming through to you and you need to go further.
6. Show the class the second page of the PDF, which has numbers grouped into houses.
7. Ask the student/group to show which house has the chosen number.
8. After deep thought, tell them the number.

The trick is that there is only one number with the given shading/colour in a given house; so long as you have the colour and house, you should be able to find the number.

Problem steps

Step 1

Either do the trick yourself or get your students to look at it online—see the series website for the link.

Go through the trick a couple of times so that the students understand the rules. Make sure that they write down their chosen number so they don't forget it along the way. Then ask them to think about why the trick works. Let them make conjectures and justify them.

As a class, discuss the trick and discover why it works. What data do they need to collect to make sensible conjectures?

In this trick there are 25 numbers, which are presented twice: once in a 5 × 5 square, picked out in one of five colours, and once in five 'houses' of five numbers each. So each number has two attributes: a colour and a house.

Each of the 25 combinations of colour and house is a unique identifier. You can thus cross-reference the combination in Table 1.6 to identify the number. For instance, 'Red/A' is 12, 'Green/D' is 21 and 'Orange/B' is 9.

Table 1.6: Colour/house combinations for the 'guess your number' trick

Houses	Colours				
	Blue	Green	Orange	Red	Yellow
House A	2	3	15	12	4
House B	7	13	9	1	23
House C	14	16	5	19	10
House D	11	21	22	8	18
House E	17	6	20	25	24

Repeat the trick until most students have discovered the secret. Don't let the first students who have found the answer tell the other students.

Step 2

Get each of your students to make up their own two-page set of numbers in colours and houses. Let them assign colours and houses to numbers in whatever way they like. In pairs, have them check that they've assigned a different colour/house combination to each number. Then get them to play the trick with each other to check that it works.

Encourage enthusiastic students to come up with other arrangements, presenting numbers not in squares or houses but in some other way that works. They might also try using any 25 numbers, not just 1–25.

Have students test their versions of the trick on students in other classes.

Step 3

Is there anything significant about 25? Could a different amount of numbers be used?

A simple extension can be made by using 36 numbers, with six colours and six houses. Students should understand this extension quickly.

Numbers 25 and 36 are both square numbers; can you make the trick with a composite amount of numbers? If you had 30 numbers could you use six colours and five houses? We show a possible assignment in Table 1.7. Ask your students to make their own version.

Table 1.7: Thirty numbers for colours and houses

Houses	Colours					
	Red	Blue	Green	Yellow	Pink	Orange
House A	1	2	3	4	5	6
House B	7	8	9	10	11	12
House C	13	14	15	16	17	18
House D	19	20	21	22	23	24
House E	25	26	27	28	29	30

Have any students made versions that are too easy to see through? Why are some versions of the same trick easier than others? Usually it's because numbers are clumped together and a sequence of consecutive numbers all have the same colour; this makes the arrangement easier to spot.

Step 4

Now we come to a more complicated trick—'Magic shuffles'. You might want to work on this over a few lessons. The trick can be accessed via the series website, but you can do it by hand with a pack of cards. It doesn't require sleight-of-hand—just attention to detail.

1. Deal out 21 cards into three columns of seven cards each. It's important to deal the cards out left to right, adding one to each column in turn, rather than dealing top to bottom and filling one column after another.

2. Ask a student to mentally choose one card from those on display, and to tell you which column it is in.
3. Collapse the columns and gather up the cards, but make certain that the pile of cards the student selected goes in the *middle* of the stack, with the other two piles above/below it.
4. Deal the cards out the same way, going left to right to create three columns with seven cards each.
5. Ask the student to look for the card they chose, and again to tell you which column it is in.
6. Collapse and gather up the cards, again making sure that the pile the student selected is in the middle of the stack.
7. Deal out the cards the same way as before, ask the student to pick the column, and gather up the cards keeping the selected pile in the middle of the stack.
8. Deal out the cards one last time. Point to the middle card of the middle column; this should be the card the student selected.

How do the three columns of seven cards hide the identity of a card so that it can be discovered based on three attributes (the three columns that the card is in)?

From what we saw in Step 1, we could code the cards using three colours and seven houses (or seven colours and three houses) and find it in two attempts. However, this trick uses a more subtle approach that is harder to see through.

When you gather up the cards and deal them out again, these cards are separated out rather than being clumped together—and because they're in the middle of the stack, they'll be distributed evenly through the middle of the new set of columns. When the student picks the correct column, that means that their chosen card must be around the middle of that column—either the third, fourth or fifth card.

When you gather up and deal out the cards again, the chosen card *must* be the exact middle (i.e. fourth) card in its column, because you situated its original pile of cards in the middle of the stack and then distributed the three middle cards evenly through the middle of the spread.

The final deal is just for show, letting you place the chosen card in the centre of the middle column.

Step 5

Listen carefully to the students' explanations of this trick to make sure they have actually solved the problem.

To confirm that they really understand what's going on, ask them to work with a smaller number of cards. Have the class make a version of this trick that uses *nine* different cards—three cards in three different columns—but that goes through the same process. This should be relatively straightforward. The class can move on to using 12 or more cards later.

Where to from here?

- What is the largest number of cards that this 3-column trick will work for?
- If you were able to use four columns, what is the largest number of cards that this trick will work for?
- Ask them to explain carefully all of the steps they used in solving these problems. What helped them to see how the trick worked?
- Ask the students:
  - What data did you collect?
  - What conjectures did you make?
  - How did you show the false ones were false? How did you justify the correct conjectures?
  - At what points along the way would you have liked help? How did you get past these points?

Level 3: Strange signs

Problem

Perform a magic trick for the class, following these steps.
1. Download the PDF teacher file for this activity from the series website.
2. Use a projector or interactive whiteboard to show the PDF (the collection of numbers and symbols) to the class.
3. Ask a student to secretly choose a 2-digit number, and to subtract the sum of the digits from that number.
4. Look away from the screen/whiteboard while the student looks for the number they just calculated. Have them remember the symbol next to their number.
5. Turn back to the screen/whiteboard, point to the ♐ symbol and tell them that this is the symbol that goes with the number they calculated.

The trick is that the calculation will always produce a number that's divisible by 9, and all of the numbers that are divisible by 9 have the same symbol.

Step 1

Get several students to try this trick until the whole class sees through it.

Step 2

Ask the students to make up a version of this trick for themselves. They should then practise it and try it on their family and friends.

Step 3

Why do they always get multiples of 9? Have them make conjectures and try to justify them.

A good way to work on this is trying the calculation with several numbers. For instance, suppose that the number they chose was 62; 62 less the digits 6 and 2 is

$$62 - 6 - 2 = 60 - 6 = 6(10 - 1) = 6 \times 9$$

The same thing happens no matter what 2-digit number you choose.

Step 4

A similar trick—'All Change'—demonstrates the same principle. The trick can be accessed via the series website, but is very easy to do in person and doesn't require any equipment. It does require talking about 1 cent coins, though, which students won't be familiar with, but they should pick up the idea quickly.
1. Tell the students to imagine they have a handful of 1 cent and 10 cent coins, but not to tell anyone how many.
2. Ask them to add up the value of their imaginary coins. From this total value, subtract the *number* of coins they imagined holding to get a new number.
3. Tell them that if this is a two- or three-digit number, add the digits together. Keep adding the digits until each student is thinking of a single-digit number.
4. Tell the students that the number they're thinking of is nine.

36 Creative Activities in Mathematics: Book 2

Let the students try the trick to see if they can work it out for themselves.

This is exactly the same trick as the one they did before, but in a different situation. The amount of money is the same as the chosen 2-digit number; the number of coins is the same as the sum of its digits. After they do the subtraction they will get a multiple of 9, so the sum of this number's digits will also be a multiple of 9. If you add up this number's digits, it will also be a multiple of 9. Eventually the answer will just be 9.

Step 5

Is there a larger number pattern here? What sort of number do you get if you take any 3-digit number and subtract the sum of the digits from it?

$$359 - 3 - 5 - 9 = 300 - 3 + 50 - 5 = 297 + 45$$

Since 297 and 45 are both divisible by 9, the difference $359 - 3 - 5 - 9$ is as well.

Can the class see that this works for any 3-digit number? Let them justify this by using an arbitrary 3-digit number.

You can then repeat the problem adding in 100 cent (1 dollar) coins. They should get the same answer of 9.

Step 6

What about 4-digit numbers? Note that $1000 - 1 = 999$; $100 - 1 = 99$ and $10 - 1 = 9$ and these are all divisible by 9.

This level is all about divisibility by 9. Talk to the class about the rule for divisibility by 9: if the sum of the digits of a number is divisible by 9, then so is the number. Let them experiment with any numbers they like to check that the rule works. Can they justify it by using the $100 - 1$ approach?

Where to from here?

- Can your students make up their own version of the 'strange signs' game starting with 3-digit numbers?
- Ask them to explain carefully all of the steps they used in solving these problems. What helped them to see how the tricks worked?

Level 4: Think of a number

Problem

Perform a magic trick for the class, following these steps; each is an instruction to the students.

1. Think of a number (x).
2. Add 5 ($x + 5$).
3. Double the results ($2x + 10$).
4. Add 40 ($2x + 50$).
5. Divide by 2 ($x + 25$).
6. Subtract the number you first thought of (25).
7. Multiply by 4 (100).

Get them to write down their first number and every number they get at each stage. When they finish, tell them that 100 is their last number.

You can see from the process above how the trick works. Make up other tricks like this with as many or as few steps as you like. You can choose numbers that students have been using in class, such as negative numbers or fractions, to give them practice in manipulating these.

Step 1

Perform the trick as shown, making sure students write down their numbers.

Unless a student makes a mistake, the answer will always be 100. If a student doesn't get a result of 100, check out the steps with them to see where their error was.

Step 2

Repeat the trick several times and record the students' answers every time.

One possible way to speed this up is by using a spreadsheet with the calculations/steps built in. That would show that the answer is 100 for every number the student can think of.

Step 3

From what they have written down, can students tell how the trick works?

The fact that they get 100 every time is not a proof but it does point towards one. More proof-like is the following list.

1. Think of 3.
2. After 3 we get $3 + 5$.
3. Then $2 \times (3 + 5) = 2 \times 3 + 2 \times 5$.
4. Next $2 \times 3 + 2 \times 5 + 40$.
5. Then $3 + 5 + 20$.
6. Now $3 + 5 + 20 - 3 = 5 + 20$, at which point the number you originally thought of disappears.
7. Finally, $4 \times (5 + 20) = 100$.

Do this with a couple of different starting numbers. Students should then see that their original number will always disappear at Step 6.

Step 4

Get students to work in pairs to invent a trick of their own like this one. Let the pairs practise their trick with other pairs. Can students work out how the other pairs' trick works?

Step 5

Now try the following sequence.

1. Think of a number (x).
2. Multiply it by 2 ($2x$).
3. Add 10 ($2x + 10$).
4. Divide by 2 ($x + 5$).
5. Subtract the number you first thought of (5).

The answer is always 5. Can they conjecture how this works? Can they justify this conjecture?

Step 6

Have each student make up one of these mind-reading tricks for themselves. Let them practise it on other students, passing teachers or at home. Can they justify the answer that they get every time?

Where to from here?

- Let them try the trick using an x or other symbol (a □ or even 'my number' will do) for their number. Does that make it easier to see how it works? You may need to help them through the steps if they have done very little algebra.
- Does it matter if the number you started with was bigger than 10? (No. The first trick probably asks you to use a number from 1 to 10 so that it is easier to do the arithmetic in your head.)
- Does it matter if the number is a fraction or decimal? (No.)
- Ask the class to write down their answers to these questions:
  - Where did you get stuck in these problems?
  - What would you have liked your teacher to tell you at this point?
  - Do you now know how to get 'unstuck' here?
  - What conjecture did you have that you can justify?
  - Can you write down a justification?
  - Does this justification convince a friend?

CHAPTER 3:
MR MAC'S iPHONE

Initial problem

Mr Mac has bought a new iPhone. Unfortunately, the battery is faulty. When he bought the iPhone, the battery had a life of 64 hours. At each charge, it ended up with twice the life of its original capacity until it got to a maximum life of 1024 hours.

How many charges did it take Mr Mac's iPhone battery to get to this maximum value?

Background information

This activity develops from a fun problem that deals with doubling through other multiples to learning about powers of 2. Mr Mac has problems based on the idea that the life of a battery changes with each charge, and changes in a particular way.

At Level 1, Mr Mac's battery doubles the length it will last with every charge it is given. The first four steps of Level 1 are available to all students. From Step 5 on, this activity is more suitable for Year 5 students and up.

At Level 2 Mrs Mac goes a step better; her battery life trebles. In order to simplify the calculations involved, we show how to use some simple spreadsheet techniques. The short lesson on spreadsheets makes Level 2 best for Year 5 students.

The other member of the Mac family has more serious problems at Level 3, as his battery's life decreases with each charge. Some Year 5 students will be able to get started on Level 3, but it is more accessible to Year 6 and 7 students. Also, while it is generally fine for students to use electronic devices for maths, it is important for them to use long division for the Level 3 activity. This will help them realise that fractions are only equivalent to certain types of decimals.

At the last Level we introduce powers of 2, their notation and how they behave so that we can tackle Mr Mac's problem from another perspective. Level 4 can be used as an introduction to indices for Year 6 students. Year 7 students should be able to cope with the whole level.

Table 1.8: Australian Curriculum content descriptions for the *Mr Mac's iPhone* activity

Activity level	Problem	Content descriptions
1	Mr Mac's iPhone	Year 5 Use estimation and rounding to check the reasonableness of answers to calculations (ACMNA099) Year 6 Add and subtract decimals, with and without digital technologies, and use estimation and rounding to check the reasonableness of answers (ACMNA128)
2	A spreadsheet approach	Year 6 Select and apply efficient mental and written strategies and appropriate digital technologies to solve problems involving all four operations with whole numbers (ACMNA123) ACMNA128 (see above)
3	Part numbers	Year 5 Compare, order and represent decimals (ACMNA105) Year 6 ACMNA128 (see above)
4	Powers of 2 and other numbers	Year 5 ACMNA105 (see above) Year 7 Investigate index notation and represent whole numbers as products of powers of prime numbers (ACMNA149)

If students want to know the real story about phone batteries and their charging habits, you might look up a website such as Apple's guide to battery life—see the series website for the link.

Big ideas
- Fractions and decimals, their use and relationship
- Rounding
- Powers
- Long division
- All fractions are decimals of a certain sort, but not all decimals are fractions

Suggested resources
- Calculator
- Microsoft Excel spreadsheet

Problem aims
- To practise the use of fractions and decimals
- To convert fractions to decimals and vice versa
- To see that fractions are the same as stopping decimals or recurring decimals

Key concepts
- Powers of two
- Rounding

Possible heuristics/strategies
- Experiment
- Keep records
- Use tables
- Draw graphs

Special notes

Powers of 2: These are numbers that are made by doubling 1 or multiplying a whole lot of 2s together.

Powers of 3: These are numbers that are made by multiplying a whole lot of 3s together.

Powers of 10: These are numbers that are made by multiplying a whole lot of 10s together.

Powers of any number: These are numbers that are made by multiplying a whole lot of the original number together.

Recurring decimals: These are decimals whose digits recur in the same way from some point on.

Level 1: Mr Mac's iPhone

Problem

Mr Mac bought a new iPhone. Unfortunately, the battery is faulty. When he bought the iPhone, the battery had a life of 64 hours. At each charge, it ended up with twice the life of its original capacity until it got to a maximum life of 1024 hours.

How many charges did it take Mr Mac's phone battery to get to this maximum value?

Problem steps

Step 1

This is simply a matter of doubling the battery life numbers until 1024 is reached. The process goes:

$$64 \rightarrow 64 \times 2 = 128 \rightarrow 128 \times 2 = 256 \rightarrow 256 \times 2 = 512 \rightarrow 512 \times 2 = 1024$$

So four charges (or doublings) are needed.

In the early stages of this activity, we suggest that students work by hand. This gets tedious as the activity develops, so let them use calculators later.

Step 2

Ask the students: is there any other length of battery life that would get to 1024 hours in four charges?

This will require them to experiment with a variety of numbers.

The answer above shows that if the battery originally had 128 hours of life, it would charge to 1024 hours in three charges. This suggests that anything bigger than 128 would need less than four charges, and anything smaller than 64 would need more than four charges.

So what about 70?

$$70 \rightarrow 70 \times 2 = 140 \rightarrow 140 \times 2 = 280 \rightarrow 280 \times 2 = 560 \rightarrow 560 \times 2 = 1120$$

But the original question tells us that Mr Mac's iPhone can't charge to more than 1024 hours. So the 70-hour battery would stop at 1024 hours, and what actually happens is:

$$70 \rightarrow 70 \times 2 = 140 \rightarrow 140 \times 2 = 280 \rightarrow 280 \times 2 = 560 \rightarrow \text{with another charge it stops at 1024}$$

So anything from 64 up to almost 128 will need four charges to get to 1024.

Step 3

Suppose we say that the *doubling number* of a battery is the number of times it has to be put on the charger to reach a life of 1024 hours.

We know that if the original life is between 64 and just less than 128, then its doubling number is 4. What if the doubling number is 7? How many hours does the battery have initially?

A little experimentation with a doubling number of 7 gives an original battery life of anything from $2^3 = 8$ hours to less than $2^4 = 16$ hours. This experimentation can be done in two ways. First, some numbers can be doubled seven times until they get to 1024 hours (or from less than 1024 to less than 2048). This might be speeded up by just considering numbers such as 16, 32, 64 (numbers that only have factors of 2) and so on to begin with.

The other option would be to start at 1024 and work back by dividing by two. Students should again find that the battery has between 8 and 16 hours of initial charge.

Step 4

What is the doubling number for the battery when it has completely run out?

This is a problem because $0 \times 2 = 0$. Mr Mac will have to get a new battery!

Discuss what your students have done over the last few steps and what methods they have used. Did they use a method other than those mentioned here? (There is another method in Level 3 and another in Level 4.) What do they think is the best method so far?

Step 5

Like all batteries, Mr Mac's battery loses power more quickly when it is being used. If his battery has a full 1024-hour charge, and he leaves it turned on but doesn't use it, it will last for 1024 hours. How many days is that?

$$\frac{1024}{24} = 42\frac{2}{3} \text{ days}$$

On the other hand, if he uses it constantly (he likes listening to music through his earphones) it will run down in 64 hours.

Mr Mac usually sleeps for 8 hours and he turns his iPhone off completely for that time. How long is it before his battery runs down?

With music on continuously, the battery would last for $\frac{64}{24} = 2\frac{2}{3}$ days. Because he sleeps for 8 hours, Mr Mac uses his iPhone for 16 hours a day. The charge will last for $\frac{64}{24} \times \frac{24}{16} = 4$ days if used this way.

Step 6

What if Mr Mac forgets to turn the iPhone off every night? Suppose he keeps the music going while he is awake, and turns the music off (but not the iPhone) during the night. Challenge students to work out how much of the battery will be used every day.

With the music off, Mr Mac uses 8 hours' worth of the battery every day just because the iPhone is on. For every hour the music is on, he uses $\frac{1024}{64}$ hours of charge. So he uses $\frac{1024}{64} \times 16 = 256$ hours for every day he has the music on. So altogether he uses $8 + 256 = 264$ hours of battery every day. This means that the battery will last $\frac{1024}{264}$ days or just under 3.9 days.

If this approach is too hard for your students, they could creep up on it using another situation. Think of the battery life as being money. If they won \$1024 in 64 hours, how much would they win in 16 hours? Now convert the problem back to battery life.

Step 7

The problems in the previous two steps are about ratios, which are often difficult for students.

You can make it more immediate by asking them to look at their own phones and tablets. How much do they use them? How long do their batteries last when they are being used and when they are not? How long would a charged battery last in their own devices if they had Mr Mac's batteries?

Where to from here?

- What maths did your students think they were doing here? (Arithmetic; doubling; ratios; fractions etc.)
- Have students invent their own problems based on the size of the battery, the amount of battery used for various purposes, and how much charge the phone used when turned on but not in use.
- Under the conditions of Steps 5 and 6, what might Mr Mac be doing with his iPhone if the battery lasted 8 days?

Level 2: A spreadsheet approach

Problem

Mrs Mac bought a new iPhone. Unfortunately, her battery is even more faulty than her husband's. When she bought it, the battery had a life of 27 hours. At each charge, it ended up with three times the life of its original capacity until it got to a maximum life of 6561 hours.

How many charges did it take Mrs Mac's phone battery to get to this maximum value?

Call this number the *trebling number* of the battery. What is the trebling number of Mrs Mac's 27-hour battery?

Problem steps

Step 1

This can be done in the same way your students did it in Level 1. However, the arithmetic is getting tedious, so this activity involves some simple uses of spreadsheets to find the answer that students should be able to understand without too much difficulty.

We use Excel throughout the directions, as this is a common program almost all schools can access. We also recommend using a projector or interactive whiteboard to work through the Excel content yourself so that students have a clear idea of what to do and how to do it.

A good first step is to go through a tutorial as a whole group or with individual students to demonstrate how to do basic arithmetic operations in Excel. Let your students practise until you are satisfied that they have mastered this. Point out that if they click on a cell, it highlights a letter across the top and a number down the side.

Step 2

Ask students to put a '0' in cell B4 and a '1' in cell B5. Put the cursor in B4 and move it across B5; this should highlight the two cells and surround them with a solid line. At the bottom of this solid line is a small square; tell students to click this square with the mouse, hold the button and drag it down several squares. When they let go of the button, numbers from 0 upwards will fill the squares, one under each other.

Have students practise dragging using various pairs of numbers. Can they explain what is happening? Can they get Excel to produce a special sequence such as 1, 5, 9, 13 etc. by dragging?

Step 3

Have students write the heading 'Number of charges' in cell B2. Let them experiment to see how to widen the columns so that headings fit exactly into the given cells.

Drag down to put the list of 0 to 10 under 'Number of charges', starting from cell B4.

Step 4

The next important thing is learning to use *formulae*.

- Tell students to type the heading 'Battery life (in hours)' in cell C2.
- Enter the number 27 in C4—the initial number of hours that Mrs Mac's iPhone battery will last.
- In C5 they need to type '= C4*3'. Click to the side of that cell and notice what has happened. What do they think the '*3' does? (Multiplies by 3.)
- Now click on C5 and drag down the box around that cell until it is level with the 10 in column B. What happens?

Get the class to experiment with formulae such as =C4*3 and report back on what they find.

Step 5

We now want to put this in the context of Mrs Mac's phone problem; so have one student read the question aloud. Then ask the students to think for a minute (use a timer) about what the question is asking them to work out. Have them pair up and share their ideas. Let volunteers share with the rest of the class what they think the question is all about and what they need to do.

Students should be able to tell you that the battery originally has 27 hours' capacity, that it will then have 81 hours and so on. Stop the discussion before they solve the first section.

Do they see that they have been here before in the spreadsheet exercise? They should know that this situation can actually be modelled with a spreadsheet.

Ask them if they can identify the two *variables* (things that will change). If necessary, lead them to 'battery life' and 'number of charges'. Show how these variables are beside each other in the spreadsheet. From the spreadsheet, ask what the trebling number of the 27-hour battery is. (It should be 5: the first place that the numbers reach 6561 or more.)

Let them create their own spreadsheets to find the trebling numbers of any number of hours they like between 5 and 50. Clarify for them that the trebling number corresponds to the first time that hours charged gets above 6561.

Step 6

Again using spreadsheets, get the class to find all initial numbers of battery hours that have a trebling number of 3. (Anything between 243 and just under 729.) Now explore other trebling numbers.

Where to from here?

- Get your students to invent and solve other problems with a spreadsheet—for example, the sum of the first 1000 numbers.
- What did your students find most interesting here? Most difficult?

Level 3: Part numbers

Problem

Mark Mac has a similar phone to his parents, but the battery loses half of its battery life every time it is charged. If it starts with a battery life of 10 hours, how long before it has less than one hour at the next charge?

Suppose we call this the *half number*. What is the half number of Mark Mac's battery?

Step 1

The half number of Mark Mac's battery is 4. Your students can find this by hand—the 'arrow' way—or with a spreadsheet, which is much more efficient.

$$10 \to 10 \times \tfrac{1}{2} = 5 \to 5 \times \tfrac{1}{2} = 2\tfrac{1}{2} \to \tfrac{5}{2} \times \tfrac{1}{2} = \tfrac{5}{4} \text{ or } 1\tfrac{1}{4} = \to \tfrac{5}{4} \times \tfrac{1}{2} = \tfrac{5}{8}$$

Remember to stop when the battery life goes below one hour.

Step 2

Now get them to discover how much charge a battery with a half number of 4 can have, using Excel. They should find that all the batteries with a life from just over 8 to exactly 16 hours have a half number of 4.

Step 3

What if Mark Mac's battery loses $\tfrac{1}{4}$ of its battery life every time it is charged? If it starts out with a battery life of 4 hours, how long before it has less than an hour left?

We'll call this the *quarter number*. What is the quarter number of Mark Mac's battery?

We show the answer in the arrow notation to point out a little subtlety:

$4 \to 4 \times \tfrac{1}{4} = 1$, so $4 - 1 = 3$ is retained. Note that 4 is reduced to $4 - 4 \times \tfrac{1}{4} = 4 \times \tfrac{3}{4} = 3$. Similarly 3 is reduced to $3 - 3 \times \tfrac{1}{4} = 3 \times \tfrac{3}{4} = \tfrac{9}{4}$.

$$\text{Then } 4 \to 3 \to \tfrac{9}{4} \to \tfrac{27}{16} \to \tfrac{81}{64} \to \tfrac{243}{218} \to \tfrac{729}{512} \to \tfrac{2187}{2024} \to \tfrac{6561}{8192}$$

Notice that if you lose $\tfrac{1}{4}$ you keep $\tfrac{3}{4}$. So counting the arrows gives a quarter number of 8.

It is preferable to do this using a spreadsheet.

Step 4

What fraction number has number ranges from exactly $\tfrac{1000}{343}$ to just less than $\tfrac{10\,000}{2401}$?

(The answer is 0.3 or $\tfrac{3}{10}$. The numbers on the edge of the ranges are all of the form $\tfrac{10}{7} \times \tfrac{10}{7} \times \ldots \times \tfrac{10}{7}$.)

Where to from here?

- Ask students to invent some problems along the lines of those in Level 1, Steps 5 to 8.
- Can they list five decimals that lie between $(\tfrac{4}{3})^5$ and $(\tfrac{4}{3})^6$? Ask if they think it is easier to find fractions or decimals between these two numbers.

Level 4: Powers of 2 and other numbers

Problem

Mr Mac bought a new iPhone. Unfortunately, the battery is faulty. When he bought the iPhone, the battery had a life of 64 hours. At each recharge, it ended up with *twice* the life of its previous capacity until it got to a maximum life of 1024 hours.

How many charges did it take Mr Mac to get to this maximum value?

Step 1

This is the same problem that we asked in Level 1, but this time we look at the question in a new way, one different from those used previously and without using a spreadsheet.

At this point we introduce *powers of 2* to the students. This gives a simpler way to look at the doublings problem we did in Level 1.

On the whiteboard write 2^1 for one doubling starting from 1, 2^2 for two doublings, 2^3 for three doublings and so on. But one doubling is just multiplying by 2, so $2^1 = 2$. And 2^2 is two doublings so that's the same as $1 \times 2 \times 2 = 4$. Then $2^3 = 1 \times 2 \times 2 \times 2 = 8$.

Step 2

Ask your students what other numbers they know that are powers of 2. If they say something like 2^5, ask them what 2^5 is equal to. They might also say 16, 32, 64 etc. Can they write these in the powers of 2 notation? ($16 = 2^4$; $32 = 2^5$; $64 = 2^6$.)

From what they have done in Step 1, can they find the x in $1024 = 2^x$?

($x = 10$ because $2^x = 2^{10} = 1024$)

This concept and notation makes it easier to write numbers the students will work with for the problem. It also helps us see the boundaries of the different doubling numbers.

Step 3

Ask your students: what is the smallest number that has a doubling number of 8?

Emphasise to them that they should try to use powers of 2. The answer is 2^2.

There is a good chance that they can't see the connection with powers of 2. Encourage them to discover the answer by direct multiplication or using Excel.

Step 4

Now do some experimentation. Get them to work out the smallest numbers that have a range of doubling numbers (see Table 1.9). What is the connection with powers of 2?

Table 1.9: The smallest numbers that are doubling numbers

Doubling number	1	2	3	4	5	6	7	8	9	10
Smallest number	2^9	2^8	2^7	2^6	2^5	2^4	2^3	2^2	2^1	1

If you add the small number in the power of 2 to the doubling number, you get 10—and 2^{10} is 1024, the number we're aiming at with all of this doubling.

Going back from 2^{10}, we get the numbers in the table by halving the appropriate number of times.

Step 5

What is the smallest number that has a doubling number of 13? We can do this by going backwards from 2^{10}.

- 2^{10} divided by 2 = $\frac{1024}{2}$ = 512 = 2^9
- 2^9 divided by 2 = $\frac{512}{2}$ = 256 = 2^8
- 2^8 divided by 2 = $\frac{256}{2}$ = 128 = 2^7
- 2^7 divided by 2 = $\frac{128}{2}$ = 64 = 2^6
- 2^6 divided by 2 = $\frac{64}{2}$ = 32 = 2^5
- 2^5 divided by 2 = $\frac{32}{2}$ = 16 = 2^4
- 2^4 divided by 2 = $\frac{16}{2}$ = 8 = 2^3
- 2^3 divided by 2 = $\frac{8}{2}$ = 4 = 2^2
- 2^2 divided by 2 = $\frac{4}{2}$ = 2 = 2^1
- 2^1 divided by 2 = $\frac{2}{2}$ = 1 = 2^0
- 2^0 divided by 2 = $\frac{1}{2}$ = (or 2^{-1})
- 2^{-1} divided by 2 = $\frac{1}{4}$ = $(\frac{1}{2})^2$ = 2^{-2}
- 2^{-2} divided by 2 = $\frac{1}{8}$ = $(\frac{1}{2})^3$ = 2^{-3}

So the doubling number of $\frac{1}{8}$ or 2^{-3} is 13. Just as dividing 2^{10} 13 times gives $(2)^{-3}$, so does 10 − 13 = −3. This is an easy way to get the smallest number with a given doubling number.

But it is also worth noting that $\frac{1}{2}$ must be 2^{-1}. What is more, $\frac{1}{4}$ is 2^{-2}; $\frac{1}{8}$ is 2^{-3}; $\frac{1}{8}$ is 2^{-4}. And so it goes on downwards, with numbers like 8 and $\frac{1}{8}$ having the same power of 2 except the fractions have a negative power: 64 = 2^6 and $\frac{1}{64}$ = 2^{-6}; 8 = 2^3 and $\frac{1}{8}$ = 2^{-3}; and so on.

Step 6

What is the largest number that has a doubling number of 13?

It is tempting to say that this is 2^{-2}, but that has a doubling number of 12. In fact, there is no such thing as the largest number with a doubling number of 13. It is as close as you can get to 2^{-2} though.

Where to from here?

- If students multiply two powers of 2 together, can they express the answer in a power of 2? (Add the small numbers.) What happens if you divide the two numbers? (Subtract the two powers.)
- What happens if you add some consecutive powers of 2? Can you express the answer in terms of a power of 2? ($1 + 2 + 2^2 + 2^3 + 2^4$, for example, equals $2^5 - 1$.)
- What problems can your students make up along the lines of the questions in this level?

PART 2: MEASUREMENT AND GEOMETRY

Part 2 presents three activities centred on the Measurement and Geometry strand.

Table 2.1: Measurement and Geometry activities

Problem	Big ideas
Hubcaps	▪ Symmetry and its various aspects: reflection, rotation, translation
Goats and wheels	▪ Circles and their use to construct unusual paths
	▪ The effect that changing something can have on something else
Boxes and boxes	▪ Connecting 3D objects with their nets and other 2D representations
	▪ Exploring 3D

Some reminders before you use these tasks in your classroom:

1. The questions in the text are ones you can ask your students. You're likely to be able to produce similar, more immediately relevant ones for your particular students as you work on these activities with them.
2. We have given suggested links to the Years in the Australian Curriculum: Mathematics for all the Levels in each activity. But given that there will be a spread of ability in your classes you should take these as a guide only. Take the opportunity to encourage every student to the edge of their comfort zone.
3. To take all students further, sometimes you can omit some of the later steps of a Level in favour of the early steps in the following Level.

CHAPTER 4: HUBCAPS

Initial problem

What is symmetry? What kinds of symmetry do students know?
Draw pictures that show symmetry of some kind.

 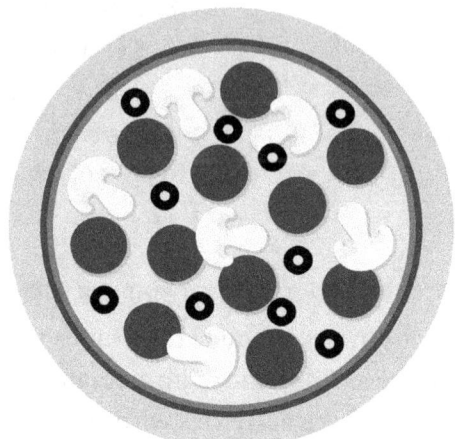

Background information

Humans like symmetry. We put it outside and inside our buildings and we use it in decorations. Our love affair with symmetry has continued for centuries, so it is natural to want to know more about it. Symmetrical objects also tend to be visually simpler in some ways and more readily described and investigated, which is one reason they have been investigated extensively in mathematics and why so much is known about them.

This activity assumes that students have encountered the concepts of line (reflective) symmetry and rotational symmetry before, but extends them and tries to increase awareness and understanding of them. The activity looks only at rotations and reflections; translations and glide symmetry do not appear.

In Level 1 we concentrate on reflections and rotations of 90° or 180°. This is appropriate for Year 4 students.

This is continued in Level 2, but we encourage students to make their own art with 30°, 45° and 60°. Level 2 can be done by most Year 4 students, and all students in Years 5 to 7 should be able to do it.

Addressing hubcaps (or rose windows) in Level 3 allows students to estimate angles and look at rotations through any angle. Some of the Level 3 work can be done by Year 4 students, but the whole activity is accessible to Years 5 to 7.

Finally in Level 4 we encourage students to design their own cars along with manufacturers' icons, fancy hubcaps and other symmetrical design features. Level 4 can be done by a wide range of students in Years 5 to 7. More depth would be expected of the older students.

Table 2.2: Australian Curriculum content descriptions for the *Hubcaps* activity

Activity level	Problem	Content descriptions
1	Pizzas	*Year 5* Describe translations, reflections and rotations of two-dimensional shapes. Identify line and rotational symmetries (ACMMG114)
2	Indigenous Australian art	*Year 5* ACMMG114 (see above) Apply the enlargement transformation to familiar two-dimensional shapes and explore the properties of the resulting image compared with the original (ACMMG115)
3	Hubcaps	*Year 5* ACMMG114 (see above) *Year 6* Investigate combinations of translations, reflections and rotations, with and without the use of digital technologies (ACMMG142) Introduce the Cartesian coordinate system using all four quadrants (ACMMG143)
4	Design a car	*Year 5* ACMMG114 (see above) *Year 6* ACMMG143 (see above) *Year 7* Describe translations, reflections in an axis, and rotations of multiples of 90° on the Cartesian plane using coordinates. Identify line and rotational symmetries (ACMMG181)

All the terms used here regarding symmetry can be found in the glossary of the Australian Curriculum: Mathematics.

Big ideas

» Symmetry and its various aspects: reflection, rotation, translation, axes of symmetry

Suggested resources

» Paper cut into circles, squares and rectangles
» Blocks
» Graph paper
» Scissors

Problem aims

» To learn about symmetries
» To demonstrate their knowledge

Key concepts
» The idea of symmetry
» Rotations, reflections, translations, axes of symmetry

Possible heuristics/strategies
» Make a drawing
» Make a model
» Trial and error

Special note
Before starting the *Indigenous Australian art* and *Hubcaps* sections, make sure that you have ready access to images of the art, and that the students have collected photos of some hubcaps.

Level 1: Pizzas

Problem

What is symmetry? What kinds of symmetry do students know?
Draw pictures that show symmetry of some kind.

Problem steps

Step 1

Have a class discussion about this. Encourage your students to see that symmetry is an action that moves a shape into the same shape in some way. Try to let them bring out the ideas of reflection, rotation and translation as well as axes of symmetry. Let them draw possible symmetrical shapes on the board and give examples from real life.

You might find it useful in class to show the YouTube video 'Rotational Symmetry, Order and Angle of Rotation'—see the website series for a link—and then discuss what it's saying. There are other videos online that can help your students here; give them the chance to find some of these.

Now ask the students to try to make up some objects that have rotational symmetry. This can be done by using Microsoft Word's 'Insert Shapes' function. They could do something similar (but more time-consuming) by carefully drawing objects on paper, tracing over them and then rotating the traced image.

Step 2

A pizza maker cuts a (circular) pizza straight through its centre. What can you say about the two pieces obtained?

Ask the class what they think can be said about the two pieces. How many ways can they make this cut?

Get one or more of your students to cut a circle of paper, representing the pizza, through its centre. (Because a circle has an infinite number of radii and each is a line of symmetry, there is essentially only one cut that goes straight through the centre of the pizza.)

The result can be found in Figure 2.1.

Figure 2.1: A straight cut through a pizza

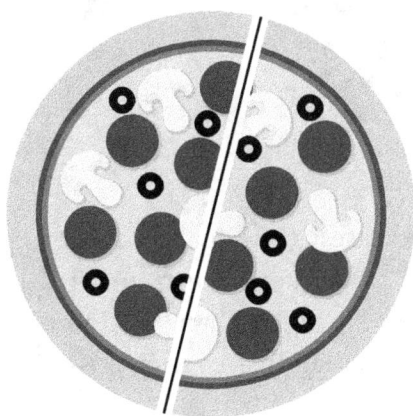

One problem here is finding the exact centre of the circle. This can be done by letting students construct the circle with a compass.

(If students haven't used a compass before, teach them that the hinge needs to be tight so that the point and the pencil parts don't slip away from each other, that the pencil needs to be placed firmly in the compass and that the pencil lead should be in line with the point.)

Another option is to create the circle digitally with its centre marked, using Word or a geometric software package, and then print it for cutting.

The single cut produces two pieces of pizza that are exactly the same. If you wanted to share a pizza equally between two people, this is the simplest way you could do it.

Note that the cut is along an axis of symmetry for the pizza. Get students to demonstrate what this, is using a circle with a line drawn where the cut is. At the same time they should show what 'axis of symmetry' means. Are there any more axes of symmetry? (One at right angles to the circle through its centre. The symmetry here is a rotation about 180°.)

Step 3

Now suppose that the pizza maker makes two straight cuts through the centre of a pizza. What can your students say about the four pieces that have been produced? How many ways do they think that this could be done?

Again, get one or more of your students to make the two cuts using a paper pizza. Discuss what happens with the class. How many different ways can the second cut be made? We show two possibilities in Figure 2.2. Is this what your students predicted?

There is an infinite number of answers, but for our purposes there are two sorts of answers: either the four new pieces are exactly the same (Situation A) or there are two pairs of two equal pieces (Situation B).

Figure 2.2: Two ways to cut a pizza

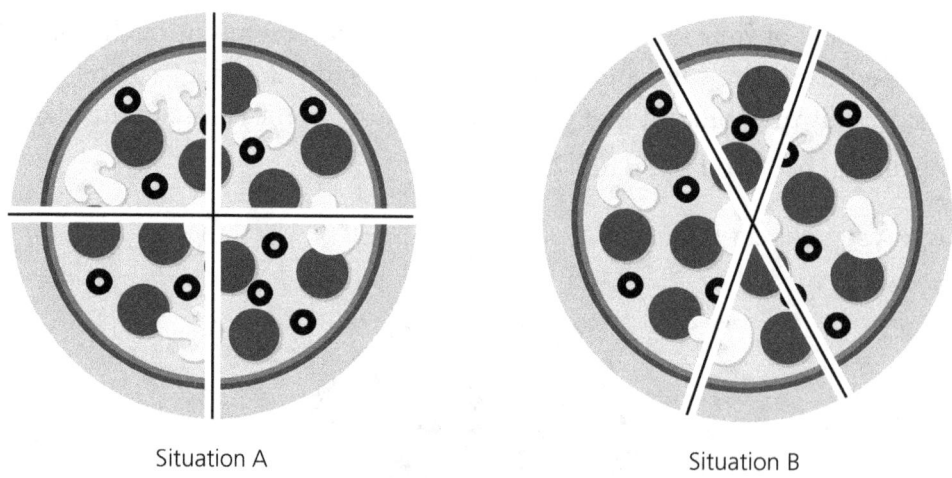

Situation A   Situation B

Why are there essentially two cases? (In Situation A the angles at the centre of the circle are all equal and equal to 90° or a quarter turn. In Situation B two of the angles in the centre are less than 90° and two are bigger than 90°. There is an infinite number of ways to produce Situation B. These two situations cover all possibilities.)

By the way, you can still divide the pizza equally between two people no matter which pair of cuts has been made. Situation B just gives each person one large and one small piece—it doesn't matter which large and which small piece.

Step 4

Symmetrically, in what other ways are the two situations different?

Let the class discuss what is going on here. Get them to think about symmetry.

Figure 2.3: Various axes of symmetry

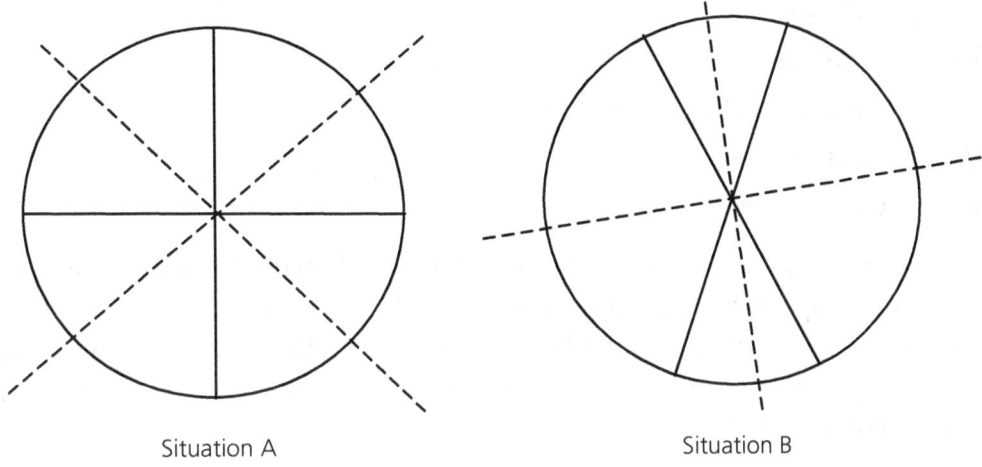

Situation A   Situation B

In Situation A there are four axes of reflection (or line symmetry). Two of these are the original cuts and two are at 45° to these cuts. The two solid and two dotted lines show the axes of reflection. In Situation B there are only two axes of reflection, shown as dotted lines.

What other types of symmetry do the cut pizzas have? Situation A has rotational symmetry about an axis through the centre of the pizza, and perpendicular to the plane of the pizza. The rotational symmetry is through a quarter turn (90°), a half turn (180°), three quarters of a turn (270°) and a full turn. On the other hand, Situation B has only a rotation through a half turn and a full turn. There are other axes for rotational symmetry.

Get the class to demonstrate the symmetry and show the axes of symmetry. They can use uncut pizzas that have the cuts drawn.

List all of the symmetries (perhaps in a table) for use in subsequent steps.

Step 5

Now provide the class with pictures of situations A and B (Figure 2.2) and ask them to colour the cut pizzas *on both sides* so that Situation A has:

1. the same symmetry as before they were coloured
2. only two reflections
3. only one reflection
4. no symmetry at all;
   and Situation B has:
5. the same symmetry as before they were coloured
6. only one reflection
7. no symmetry at all.

Get the class to report back when they're finished, showing what they've done and how it satisfies the symmetry conditions. Note that some of these symmetry conditions can be achieved in more than one way.

Step 6

Is it possible to colour situations A and B on just one side so that Situation A has:

1. the same symmetry as before they were coloured
2. only two reflections
3. only one reflection
4. no symmetry at all;
   and Situation B has:
5. the same symmetry as before they were coloured
6. only one reflection
7. no symmetry at all?

Get the class to report back when they are finished, showing what they have done and how it satisfies the symmetry conditions. For example, in 1 and 5 any rotational symmetry about the axis perpendicular to the plane of the pizzas will not be disturbed. Note that some of these symmetry conditions can be achieved in more than one way, while some might not be possible at all.

Step 7

Is it possible to put circular 'olives' on the pizzas so that Situation A has:

1. the same symmetry as before they were coloured
2. only two reflections
3. only one reflection
4. no symmetry at all;
   and Situation B has:
5. the same symmetry as before they were coloured
6. only one reflection
7. no symmetry at all?

Get the class to report back when they are finished, showing what they have done and how it satisfies the symmetry conditions. Note that some of these symmetry conditions can be achieved in more than one way, while some might not be possible at all.

Step 8

Not all pizzas are circular; some are square. Does the change of shape alter anything that we have done up to now?

Let your students work with square shapes to do their cutting and consider their symmetry, their axes of symmetry and their colouring symmetry. When this work is complete, have a class discussion on what they found. This discussion should include their reasons for their conclusions and demonstrations of any symmetry.

Step 9

Some pizzas are rectangles that are not square. Does the change of shape alter anything that we have done up to now?

Let your students work with rectangular shapes to do their cutting and consider their symmetry, their axes of symmetry and their colouring symmetry. When this work is complete, have a class discussion on what they found. This discussion should include their reasons for their conclusions and demonstrations of any symmetry.

Where to from here?

- What other shapes have the same two-cuts pizza properties as the circle and the square? Why? (Regular hexagons and regular octagons.)
- What other shapes have the same two-cuts pizza properties as the rectangle? Why?
- What other shapes have different two-cuts pizza properties to the shapes above? Why?
- What symmetries still exist if the pizzas are cut *three* times through the centre?
- What other ideas about symmetry do your students have?

Level 2: Indigenous Australian art

Problem

Here is a piece of Indigenous Australian art by an Aboriginal artist.

What can your students find out about this piece of art?

Figure 2.4: Aboriginal sand/dot painting

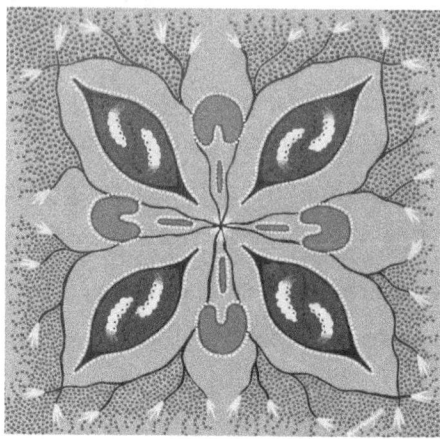

Artist: Aunty Esther Quinlin, *Dhubal in Bark*

 A file with a colour version of this image can be downloaded from the series website.

Problem steps

Step 1

The first step, before getting into the mathematics, is for students to see the significance of sand/dot art like this within Indigenous Australian cultures. Get the students to consider the art as a ritual and try to understand the significance of each part of the art. It is important to also find out why and where the piece was made, and who made it. This piece, *Dhubal in Bark*, is by an Aboriginal artist named Aunty Esther Quinlin; encourage students to search online for more information about her and the kind of artworks she creates.

 Websites such as Indigenous in Style or DNAAG—see the series website for links—are worth consulting, but it is better still to have someone knowledgeable talk to your students. If possible, make contact with an Indigenous gallery curator or Indigenous artist in your area, perhaps with the assistance of the school or local government resources, and see whether they will visit the class to talk about their work and the importance of art in their community. Any Indigenous students in your class might also be able to provide their own perspectives.

Step 2

Now get your students to look at the art from a mathematical perspective. What symmetry do your students see in this piece of art? Get them to look very carefully.

At first sight there seems to be quite a bit of reflective and rotational symmetry—but in fact there is very little symmetry here. The subtle positioning of the 'plant' motifs that extend from the centre means there is no symmetry at all.

Remind the class that symmetry in art may not be exact. The work is produced by people, not machines, so there's likely to be some variation from a mathematical ideal of symmetry. Artists may also not want their work to be exactly symmetrical. This needs to be considered when examining art for symmetry.

Step 3

What symmetry can your students find if the plants are removed from the picture? Does this reduce the symmetry or increase it?

This increases the symmetry. There are now reflections or line symmetries about a vertical axis and a horizontal axis.

Ask the students what they think this change would do to the significance of the art.

Step 4

What happens if the witchetty grubs are removed? Is it always the case that removing aspects of a symmetrical object increases the amount of symmetry?

Sometimes it makes no change in the symmetry—see Figure 2.5(a). Can your students give an example of this? Sometimes you can actually *add* symmetry—see Figure 2.5(b).

Figure 2.5: Changing symmetry

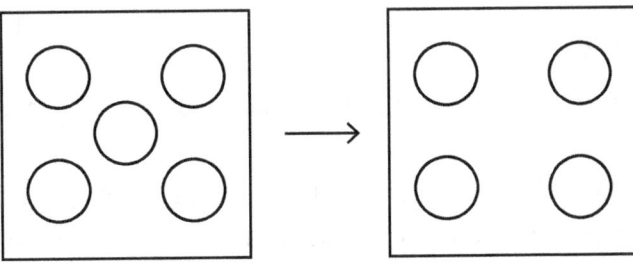

Figure 2.5(a) No change in symmetry

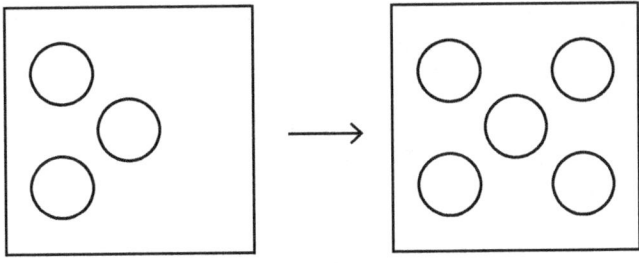

Figure 2.5(b) Increase in symmetry

Step 5

Note that the piece of art in Figure 2.4 is square. Can a rectangle have all of the symmetries of a square?

(No: it can't have a reflective symmetry through opposite corners and can't have rotational symmetry through 90°.)

Step 6

Let your students find a dot painting that has some symmetry. This is probably most easily done by searching online, though your school library may also have some books on Indigenous Australian art.

Encourage them to discover all of the symmetries of the art they have chosen. Give some students the opportunity to discuss the works that they have chosen and to show how the symmetry works.

Do dot paintings ever have no symmetry?

Step 7

At this point your students can make their own drawings using dots. (Before doing so, it is important to explain that Aboriginal paintings have important spiritual and cultural significance and that non-Aboriginal people should not try to paint like Aboriginal people. Instead, they should use Aunty Esther's painting to inspire them to create their own drawings and paintings made up of dots.) When they have finished, let some of them explain the symmetry in their works.

Next, choose a specific form of symmetry and let them make dot art that has this symmetry. For instance, you might want them to produce something that has two axes of line symmetry and one of rotational symmetry.

Where to from here?

- What other objects that have been made or occur naturally have line or rotational symmetry?
- Your students might like to investigate art that is on paper shapes that are not rectangles. Can new symmetries be found in these shapes that even squares do not have?
- What is the hardest symmetry to find in a piece of art?

Level 3: Hubcaps

Problem

What symmetries can be found in hubcaps?

Problem steps

As an alternative to hubcaps, you could instead examine rose windows in this activity if that's more appropriate for your school or students. Using rose windows requires only minor changes to the material.

Every student should have a picture of at least one hubcap (or rose window) when you start this activity. A day or two before you plan to start, tell them to collect images—either taking photos themselves or looking for images online—and bring them to class.

Step 1

Place the students into groups, and ask them to compare their hubcap to those of the other students. What do their hubcaps have in common?

Discuss what they have found in mathematical terms. Hubcaps are round, have rotational and line symmetry, have a manufacturer's logo in the middle etc.

(Note that you will have to ignore the manufacturer's logo when discussing symmetry. Occasionally you will also have to ignore a hole left for the tyre to be inflated.)

Step 2

Hubcaps have scope for symmetry that is not present in a square. Look at the hubcap illustration in Figure 2.6; this has rotational symmetry that a square can never have. It can be rotated through an angle that is a fifth of 360° and the hubcap seems to be in its original position—an angle of rotation of 72°.

Because it takes five 72° rotations to get the hubcap back to its original position, it can be said to have *rotational symmetry of order five*. Show this by rotating a picture of this hubcap, or an actual one that you have borrowed from someone's car.

The hubcap also has reflective symmetry. There are five axes of symmetry through the five spokes, one of which is shown in Figure 2.6.

Figure 2.6: A hubcap with pentagonal symmetry

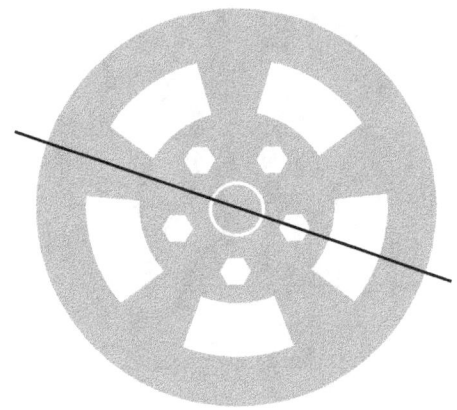

The symmetry of this hubcap is called *pentagonal symmetry* because it is the same as a regular pentagon. Ask the class if any of the hubcaps in their images have pentagonal symmetry.

Step 3

Other hubcaps may have *hexagonal symmetry* (like a hexagon), *heptagonal symmetry* (like a regular seven-sided figure), *octagonal symmetry* or even *dodecagonal symmetry* (like a 12-sided figure). Show students some examples from the images they brought to class.

Look carefully at the hexagon in Figure 2.7. Show how it can be rotated onto itself by successively turning it through 360° ÷ 6 = 60° about its centre. What order of symmetry do students think this sort of hubcap has? (It has six.)

Now show them the six axes of symmetry, one of which is drawn in the figure.

Figure 2.7: A regular hexagon

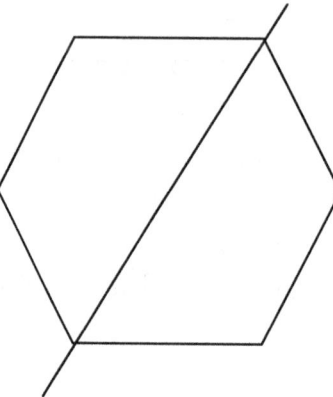

Does anyone have a hubcap with hexagonal symmetry? Let them reason why theirs does or does not have hexagonal symmetry.

Have students look at each hubcap image in their group, determine what kinds of reflective and rotational symmetry it has and compare it to different regular polygons. Make sure that the class understands the different symmetries involved and the angles needed to produce the rotational symmetries.

Note that not all hubcaps will have such simple designs as the one in Figure 2.6. The hubcap illustration in Figure 2.8 has double spokes but still only has pentagonal symmetry.

Figure 2.8: More pentagonal symmetry

What order of symmetry do your students' hubcaps have? Do any have order 1, 2 or 3? Can they explain this?

(Order 1 means no rotational symmetry and car manufacturers probably like to have symmetric hubcaps. Orders 2 and 3 have very limited rotational symmetry, which again is a problem if you are designing hubcaps for cars.)

Step 4

Ask your students to copy their hubcaps onto paper, and to draw on the copy the various angles that produce rotational symmetry as well as the axes of line symmetry. They might then paint it in two or more colours. Does this painting preserve the symmetry?

Step 5

Ask the class to individually design and draw a hubcap with pentagonal symmetry—one more complicated than the one in Figure 2.6 or even Figure 2.8.

Step 6

Many other objects have a lot of circular symmetry. Ask students to find three of these. What order of symmetry do your three objects have?

(One you might notice is the way that individual coloured lights in a traffic light are made up with small lights.)

Where to from here?

- Are there any manufacturers' logos that are symmetric? What kinds of symmetry do they have?
- Can the class find examples of rose windows (or any other symmetric windows)? Why do they think that rose windows were, at one stage, particularly popular? What angles of symmetry can rose windows have?
- What is the difference between a rose window, a wheel window and an oculus? Can they find examples to show this difference? What symmetries do wheel windows and oculi have?
- What ideas do your students have for extending the ideas in this level?

Level 4: Design a car

Problem

Your students are about to start a new elite sports car business. They have to design the prototype, and it must be as symmetric as possible. What will their car look like?

Problem steps

Step 1

Get the students into groups. It is best to take this a step at a time, or for different members of the group to design different parts of the car. The three essential parts are body, logo and tyres.

First, get them to think about the body. Before they start they should do some research on sports cars and what they look like. This may help them to get design ideas. What views of the car are important? (Front, side and from the top.) What symmetry is possible? What cars look the best? Why?

Allow groups to report their design to the whole class. They should say where they have included symmetry (or not) and why they have chosen the symmetry they have.

Step 2

Next is the logo. Each group needs to make up a name for their company and then position that within a symmetrical design. It's again worth looking at car manufacturers to see what logos they have, how symmetric they are, how complicated they are and so on. Which logo do your students think is the most appealing?

Get them to realise that they need to put dimensions on their logo design so that it can be reproduced in the factory.

They should mock up an accurate copy of their design that includes colour.

Allow some groups to report on their logo and talk about its symmetry and why they chose that symmetry.

Step 3

The final step is designing the tyres and hubcaps. Emphasise again that you are looking for symmetry, that they need to show dimensions and that they need to produce a scale mock-up of their hubcaps (with logo) for factory reproduction.

Get each group to give a presentation to the class that covers all areas of their design and shows that they know about rotational and line symmetry.

Where to from here?

- What notice of symmetry do you think that car manufacturers actually take into consideration?
- Apart from the car industry, what other industries are conscious of symmetry?
- What symmetry problems can your class create?

CHAPTER 5:
GOATS AND WHEELS

Initial problem

Gantug and his family have given up hunting and gathering and now live on a quiet fertile plain next to a river. They have just learned how to make ropes and find them very useful in stopping their goats from wandering away. Fred, their billy goat, is tethered by a rope to a stake on the plain.

What shape does Fred, the goat, wander over at the limits of his rope? What is the shape formed in the grass by the goat as he eats his food for the day?

Background information

It has been important, for many reasons over many years, to know the path that a constrained point will take. For example, if you know the *locus* (trajectory) that your cannonball will take, your aim will be considerably improved. Mathematicians have known for a long time what the trajectory is of other points under certain circumstances.

This activity is based around circles and the way that objects behave (the trajectories they take) when they are constrained to move by rotating about some fixed point. Much of it will be based on experimentation, so that students get a rough idea of what actually happens. This 'rough' picture can be later formalised to get a true picture.

We look at the attempts by a prehistoric man to tether his goat and to develop a useful wheel. In Level 1 we look at what shapes goats can make while they graze if they are tethered at one point. This level should be accessible to students from Year 4 up.

From there we look at how an ancient inventor might have developed a useful wheel by starting with triangular wheels (Level 2), then square wheels (Level 3) and finally round wheels (Level 4). At each Level we consider what happens to points at different parts of the wheel. What shapes do they trace as the wheels roll?

More able students from Year 4 can do Level 2, along with students above Year 4. Level 3 can be done by students in Years 5 to 6. A lot of Level 4 is good for able Year 5 students, but the difficulty is increasing and only students in Years 6 and 7 are likely to complete all of the activity.

Table 2.3: Australian Curriculum content descriptions for the *Goats and wheels* activity

Activity level	Problem	Content descriptions
1	Gantug's goat	*Year 4* Use scaled instruments to measure and compare lengths, masses, capacities and temperatures (ACMMG084) Compare and describe two-dimensional shapes that result from combining and splitting common shapes, with and without the use of digital technologies (ACMMG088) Create symmetrical patterns, pictures and shapes with and without digital technologies (ACMMG091)
2	Triangular wheels	*Year 4* (ACMMG088) (see above) (ACMMG091) (see above) *Year 5* Describe translations, reflections and rotations of two-dimensional shapes. Identify line and rotational symmetries (ACMMG114) *Year 6* Investigate, with and without digital technologies, angles on a straight line, angles at a point and vertically opposite angles. Use results to find unknown angles (ACMMG141)
3	Square wheels	*Year 5* ACMMG114 (see above) *Year 6* Investigate combinations of translations, reflections and rotations, with and without the use of digital technologies (ACMMG142)
4	Circular wheels	*Year 6* ACMMG142 (see above) *Year 7* Describe translations, reflections in an axis, and rotations of multiples of 90° on the Cartesian plane using coordinates. Identify line and rotational symmetries (ACMMG181)

Big ideas

» Shapes (circles, triangles and squares)

» Rotation

» That as one thing changes so does another—this provides a valuable foundation to the important concept of 'function' later in school

Suggested resources

» Two 1 × 1 squares, a 1 × 2 rectangle, a 1 × 3 rectangle and a 2 × 2 square, all made from cardboard

» Ruler and compass

Problem aims

» To use circles and parts of circles

» To see how to construct the locus of a point

» To discover symmetry in certain loci

» To see patterns in a series of geometric objects

Key concepts

» Square

» Circle

» Arcs of a circle

Possible heuristics/strategies

» Trial and error

» Use a model

» Draw a picture

Suggested resources

» Materials for making the *treel* and *squeel* carts: straws, bamboo skewers, cardboard

Special notes

Locus or **trajectory:** A locus or trajectory is the path that a point takes when it is subject to some constraint. For instance, the locus of a point that can only move so that it is a fixed distance from a given point will be a circle.

Tangent: A tangent in this context is a line that just touches a circle.

Level 1: Gantug's goats

Gantug and his family have given up hunting and gathering and now live on a quiet fertile plain next to a river. They have just learned how to make ropes and find them very useful in stopping their goats from wandering away. Fred, their billy goat, is tethered by a rope to a stake on the plain.

What shape does Fred, the goat, wander over at the limits of his rope? What is the shape formed in the grass by the goat as he eats his food for the day?

Problem steps

Step 1

The shape, shown in Figure 2.9, is a disc. The boundary of the disc, or *locus* of the tight rope, is a circle. With the rope tight, the radius is as far as Fred's neck can get from the peg.

Figure 2.9: Fred's circular boundary

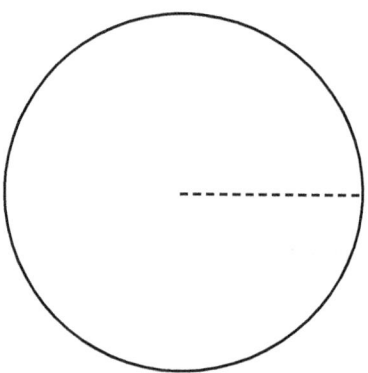

Discuss with the class how to draw a circle. They may know about drawing a circle with a compass, but it is also possible to draw a circle using a length of string. This could be done in the playground using a string with a piece of chalk at one end and a knot at the other. If one student holds the knot still at one point, and another student holds the chalk and moves so that the string is always tight, they can draw a reasonable circle.

The hard part of this is holding the knotted end fixed in one place. It might be easier if you tie the string to a fixed post or stick that becomes the centre of a circle. This could also be done in the classroom, with students holding arms outstretched, one hand representing the stake and the other, Fred's mouth.

Step 2

In the four situations shown in Figures 2.10–2.13, where the dotted line represents a rail, ask the class to predict the shape that Fred's grazing makes in the grass; then ask then to do it practically either using string or outstretched arms. Have a discussion about the results they got. What does each shape look like?

What if Fred's rope was attached to a rail that went right round a circular wall? (See Figure 2.10(b)—an annulus, although it isn't necessary to use this word.)

Figure 2.10: A new boundary for Fred

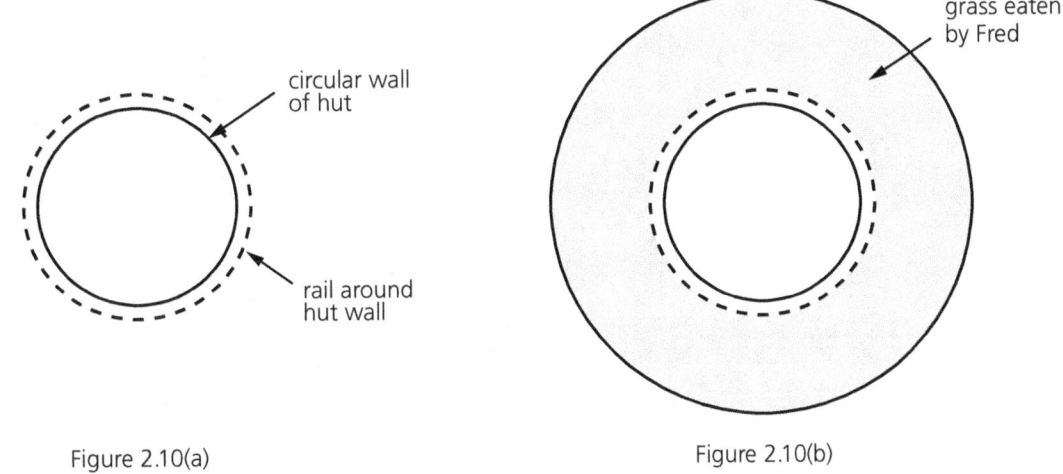

Figure 2.10(a)

Figure 2.10(b)

Get your students to draw the shape or try creating it in the playground. Explain carefully what each part of it looks like.

What shape would Fred's wanderings look like, if his rope was attached to a rail on part of the side of a straight wall? (See Figure 2.11.)

Figure 2.11: Another boundary for Fred

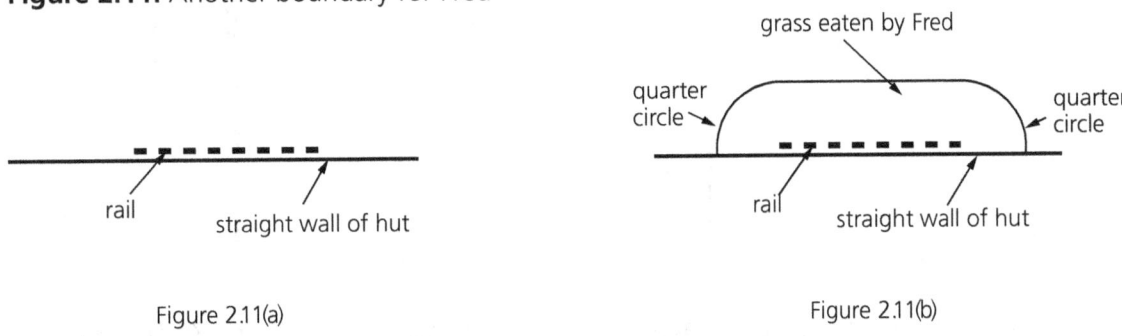

Figure 2.11(a)

Figure 2.11(b)

Gantug has made a rail along parts of two straight walls that are joined at right angles to make a corner. The rail goes into the corner made by the walls. What shapes can Fred make as he grazes? (See Figure 2.12(a) and (b) for the two possible situations. Let the students discover the ambiguity of the question. The boundaries of the shapes will consist of two straight lines and two or three quarter circles, respectively.)

Figure 2.12: More boundaries for Fred; locus along a rail next to (a) an internal corner and (b) an external corner

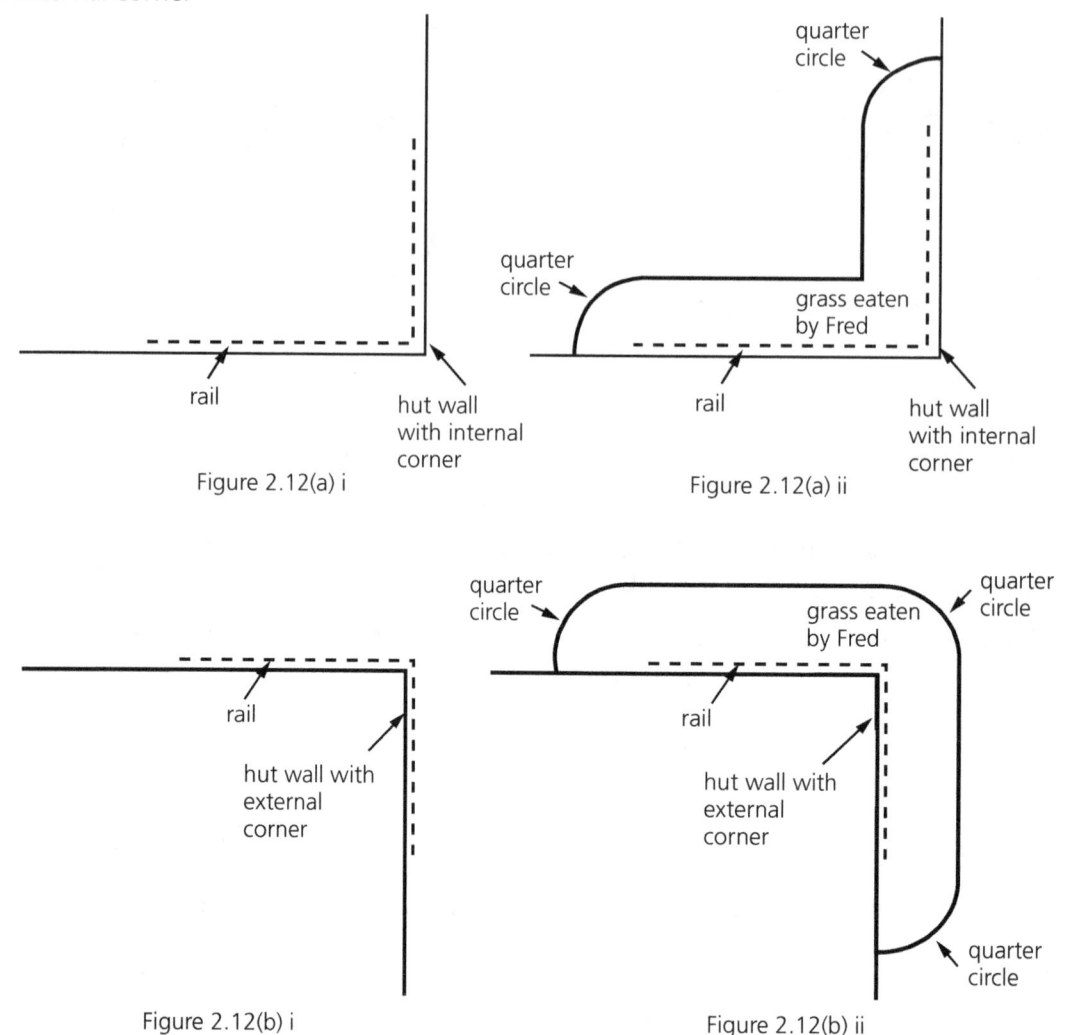

Figure 2.12(a) i

Figure 2.12(a) ii

Figure 2.12(b) i

Figure 2.12(b) ii

What if the distance Fred can reach is longer than each part of the rail? (There will be an intersection of the circular arc caused by the corner of the walls and one or more of the other quarter circles.)

Then Gantug ties Fred's rope around a short rail in a field. What shape do we get now? (See Figure 2.13(a) and (b).) Make sure that they can explain each part carefully. (Pieces parallel to the rail and up to the perpendicular to end of the rail are straight lines. At each end there is a semicircle. So the shape is a combination of two rectangles and two semi-circular discs.)

Figure 2.13: A new boundary for Fred around a short rail in a field

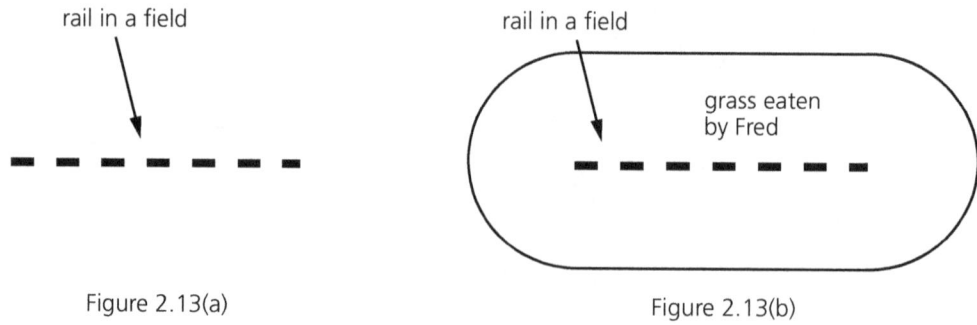

Figure 2.13(a)

Figure 2.13(b)

The trajectory or locus consists of two straight, parallel lines with two semicircular ends as shown in Figure 2.13(b). The two ends of the taut rope allow Fred to swing around much as he did when his rope was tied to one spot.

Let them suggest other ways that Fred might be tethered and find and describe the shapes of eaten grass and the boundaries of the shapes. What if there is a rail on the inside or outside of a square wall?

Step 3

Fred could defend himself from most things, but Freda the nanny goat was more vulnerable, so Gantug used a peg to tie her on a rope to a spot near his square hut. Think about the sort of trajectories that Freda could eat out as she moves on the end of her rope. Can the trajectory ever be a circle? If so, what can be said about the length of the rope? What can be said about the positions of the peg?

To make this circle, the length of the rope can never be bigger than the distance of the peg from the hut. If it were any bigger it would start to wrap itself around the building as Freda moved (see Step 4). So the only way to produce a locus of a circle would be to have the goat's mouth never reach or just touch the hut when the rope was at full stretch.

If Freda can just reach the hut, can the students fully describe where the stake has to be? This time we want to find the locus of the peg. From what we have just said, for a given length of rope, the stake has to be on the path in Figure 2.14. This has straight sides until the sides of the hut are no longer tangents to the goat's circles, and then quarters of circles on the 'corners'. The radii of these circles is again the length of the rope (plus the reach of Freda's nose).

So first, what points are *exactly* a rope's length away from the building? The easiest ones to see are the straight vertical and horizontal lines that run parallel to the hut (see Figure 2.14). But there are other points between these lines, which are a rope's length from the corners of the hut. Because these points are a fixed length from a fixed point, they must be parts of circles.

From this information, for a given length of rope, the peg has to be positioned on the path in Figure 2.14. (Points X and Y are both examples of where the peg might be placed.) This path has straight sides parallel to the hut, with a quarter circle on each corner. The radius of each quarter circle is the length of the rope (plus the reach of Freda's nose).

Finally then, if Gantug wants to let Freda move in circles, he must put the peg holding the rope on the trajectory in Figure 2.14 or anywhere outside (away from the hut) of that trajectory.

Figure 2.14: Where the peg can be placed for a fixed length of rope

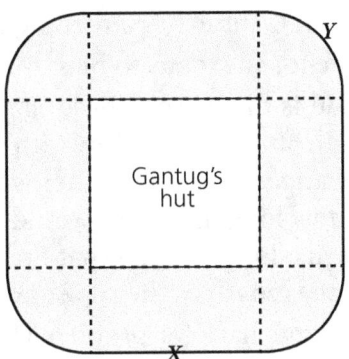

Step 4

Now put the peg for Freda's rope anywhere you like so that she can reach the hut. As Gantug makes the length of the rope longer and longer, what different shapes might Freda eat out?

No matter where Gantug puts the peg, Freda will start to produce part of a circle until the rope is long enough to let her get to the hut (see Figure 2.15(a)).

As the rope gets longer, the fun begins because Freda's next locus is a circle with parts of the hut intruding on Freda's potential circle on two sides of the hut (Figure 2.15(b)). It starts off with a radius of R_1. However, once her rope gets to the corner of the hut, it is constrained to move in a circle with the centre at that corner and with a smaller radius, R_2.

In Figure 2.15(c), the rope is even longer and is constrained by four corners of the hut, two on each side. This causes, first, part-circles of radius R_2, centred on the corners of the hut C_1 and C_2. As the rope gets longer, it gets to the other two corners of the hut and then swings round those in part-circles of radius R_3.

As the rope gets longer still, the locus gradually gets more and more complicated as the circular arcs start to cross over each other. It gets even more complicated if Freda isn't tethered symmetrically between two corners. Let the students experiment.

Figure 2.15: Part-circles drawn close to the square (hut)

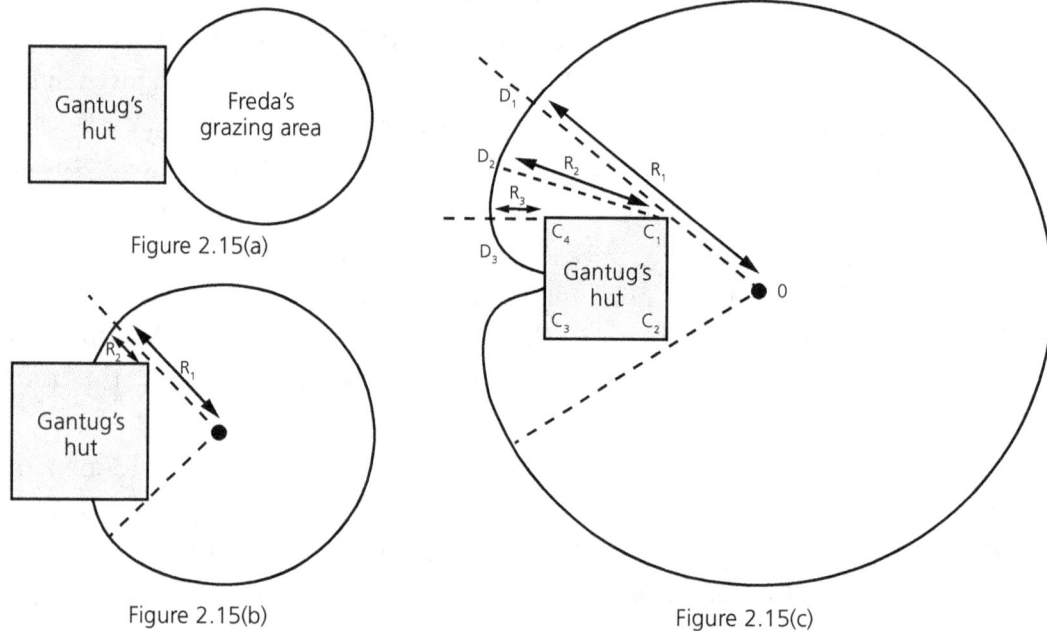

Figure 2.15(a)

Figure 2.15(b)

Figure 2.15(c)

Get students to sketch their ideas on paper, or create diagrams using the box/string setup, and put the most interesting of these shapes on the classroom wall.

Where to from here?

- What happens to Freda's locus if Gantug has a rectangular or a circular hut?
- Where does the peg have to be for the locus to be symmetric in all of these cases?
- Gantug says that his square hut is three breadths long. (Tell students that there were no international units of length at his time. A 'breadth' is the distance from the fingertips on one hand to the fingertips on the other when Gantug's arms are at full stretch. If you were Gantug, how long in metres would your 'breadth' be?) He has set up the peg so that the rope is just long enough that Freda can just nibble all around the edges of the hut. What is the length of the rope and where is the peg?
- What other ideas do your students have for problems like this? (Quadrilateral dwellings, polygonal dwellings etc.)

Level 2: Triangular wheels

Problem

For centuries Gantug's ancestors have been dragging loads over large and small distances. This was hard work and slow, even if they used goats to pull the loads. Gantug was sitting under a tree one day and thinking about this problem. Watching his daughter play with a stick on top of a stone, he was struck by inspiration and realised that he could make a *treel* (see Figure 2.16).

Figure 2.16: The invention of the *treel* and *treel* cart

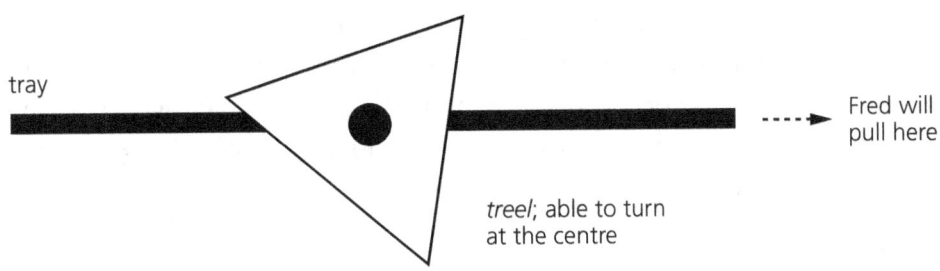

He would make a *treel*, a three-sided stone thing, and attach two of them to a big piece of wood to put the load on. Then he'd get Fred to pull it along. The *treels* would go round as the goat pulled. This would solve the resistance problem. To make life easier he would make sure that each *treel* had three equal sides.

Suppose that the *treel* cart is pulled across flat land. What is the trajectory or locus of the centre of the *treel* as the contraption moves on a flat surface?

Problem steps

Step 1

Discuss the answer with your class, then let them work out how to draw their suggested locus. We suggest that they first engage in some practical work to get a feel for what is going on. The practical work might be to make a *treel* cart and pull it along to see what happened to the centre of the *treel*. They could even make a film of the *treel* moving. A *treel* cart can be made by putting a stick through a straw and then fixing a cardboard *treel* to each end of the stick. The *treels* should be fixed so that their bases start on the ground together. The straw can then be attached to a cardboard tray.

Step 2

When they have some idea of the locus of the centre of a *treel*, get the class together to discuss their conclusions. This really is all about parts of circles.

As the *treel* moves forward, the centre itself rotates about the vertex that is on the ground. This means that its centre has to move on a series of arcs of a circle. A new arc starts whenever the next vertex hits the ground. The radius of the arc is the distance between the centre and a vertex. The locus of the centre is shown in Figure 2.17.

Figure 2.17: The locus of the centre of the *treel*

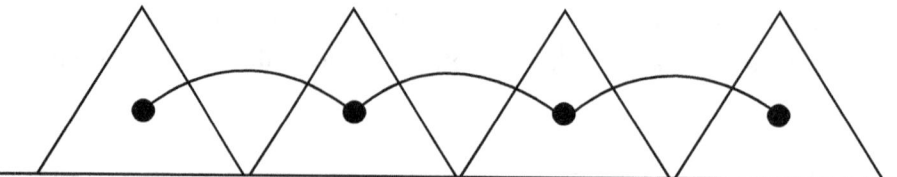

Step 3

But can we see more clearly why this works? We will do this in steps, rotation-by-rotation as in Figure 2.18. This is a bit like making a cartoon film. After every rotation we can analyse what has happened. The last position of the *treel* is shown in black and all previous positions are in grey. The current contribution to the locus is in black too and the previous contributions are in grey.

Figure 2.18: Rotation-by-rotation moves

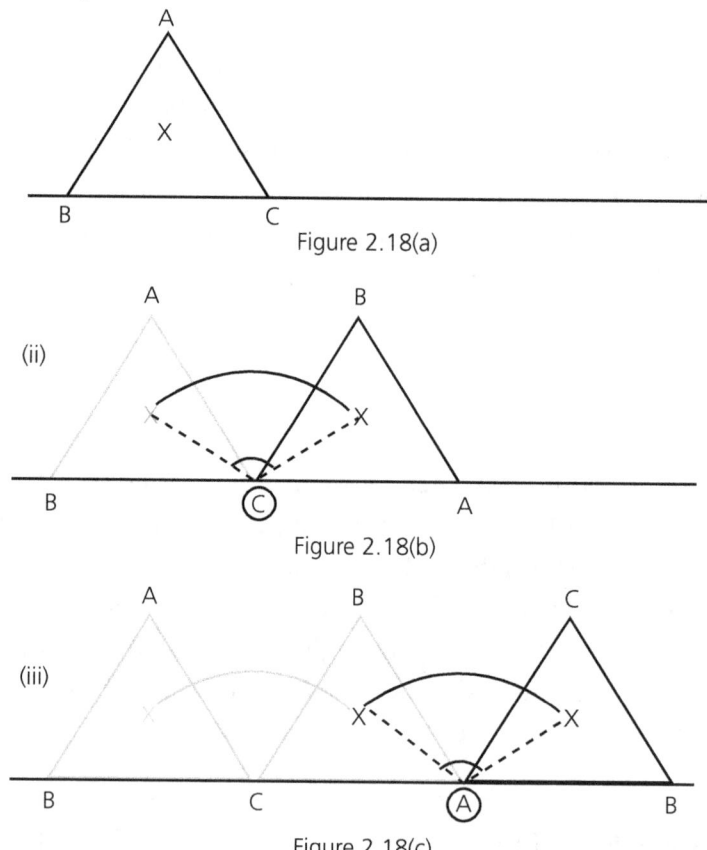

Figure 2.18(a)

Figure 2.18(b)

Figure 2.18(c)

As the *treel* turns, first one side is on the ground and then the next. The *treel* rotates through 120° in this process, so the arc subtends an angle of 120° at its centre. This should give a very bumpy ride for any load.

Position (a) is the initial position of the *treel*. To help see what is happening, the corners are labelled A, B and C and the centre of the *treel* is X.

76   Creative Activities in Mathematics: Book 2

Position (b) shows where the *treel* is after one 120° rotation. We put a ring round C because it is the point about which the centre of the *treel* is rotating. A and B have moved as shown. Here X has moved through 120° to the black X, keeping a fixed distance from C. This gives us the first part of the locus of X.

Position (c) shows what has happened in the next rotation. A is ringed here as it is the point about which everything rotates. B and C have moved as shown. Throughout the rotation, X stays the same distance from A to give us the next part of the locus. Since X is at the centre, it is the same distance from C as it is from A. So the new part of the locus looks exactly like the first part.

After this it should be clear that we get the same arc again and again.

Give the students a chance to use the rotation-by-rotation method to draw a more accurate diagram of the locus of the centre of a *treel*. Can they now make a better film of the *treel* cart moving than they had before?

Step 4

Despite the bumpiness of the cart, there was incredible joy in Gantug's clan with the invention of the *treel*. It would make it much easier to move people when they next needed to migrate. They were so happy that they put burning flares onto the vertices of the *treel*. What loci did theses flares make? Do all of the flares make the same locus?

Students might make a *treel* with cardboard folded into the shape of a triangular prism with a pencil attached at a vertex. The triangular prism doesn't have to be accurate; just fold it toughly into thirds and firm up the creases. Roll the prism around and make out the locus.

Let the class work in groups with this problem and report back. In class discussion, make sure that they describe each part of the loci in terms of circles, radii and translations (the locus of one flare is a translation of the other).

Step 5

Once the students have experimented, get them to make an accurate drawing of the loci. This should consist of arcs of circles that have radii equal to the side length of the *treel* and turn through an angle of 120°. We show this in Figure 2.19.

Figure 2.19: The loci of the vertices of the *treel*

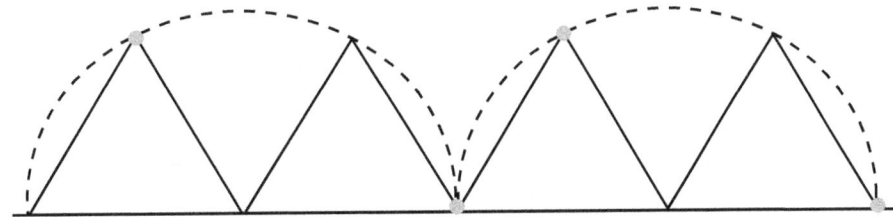

Step 6

Ask students if the loci go through any positions of the other flares. Roughly where does this happen? (It will be vertically below the vertex that has no flare and equidistant from the other two vertices that the non-flared vertex is on.)

What are the similarities and differences between the locus of a centre of a *treel* and the locus of a flare? (The angles of the arcs are the same in both cases, so some parts of the loci are concentric. The arcs for the flares are longer than those for the centre. The flares go further.)

Where to from here?

- What suggestions do your students have for more loci?
- How can the bumpiness of the *treel* be reduced? Can the ride be improved by using an isosceles triangle? What are the loci of the vertices of an isosceles *treel*? Are they still each a translation of each other?
- What if the land over which they were pulling the *treel* cart went uphill and then downhill?
- What can Gantug invent next?

Level 3: Square wheels

Problem

Of course, it had to happen. Gantug went on to invent a square *treel* (Figure 2.20). He called the square wheel a *squeel*. He made sure that each side had the length of his arm. What advantages did this have over the *treel*?

Figure 2.20: The invention of the *squeel* and the *squeel* cart

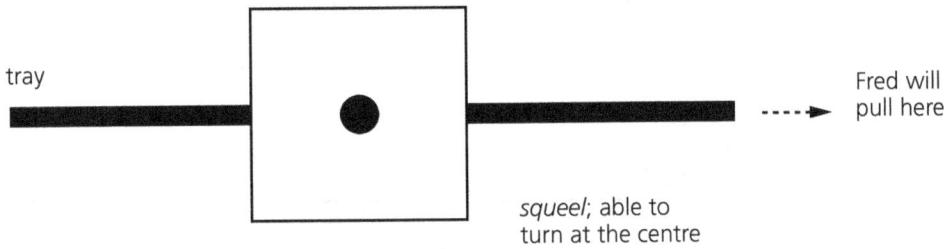

Problem steps

Step 1

One advantage is that the *squeel* gives less of a bump. Get the students to see this by experimenting with the *squeel* cart, perhaps using physical models or by using the more systematic approach of looking for arcs of circles—in particular, the arcs relating to the centre of the *squeel*. So what is the locus of the centre of the *squeel*?

Step 2

We show the locus of the centre in Figure 2.21. This can be produced by experiment or by the rotation-by-rotation method. Compare that with the locus of the centre of a *treel* when the distances between the vertices and centres of *squeels* and *treels* are the same. Because the *treel* goes through a larger arc, its centre goes up higher than the *squeel's*.

Figure 2.21: The locus of the centre of the *squeel*

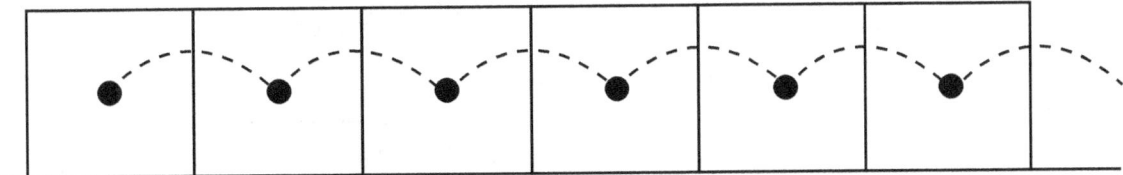

The *squeel* was less bumpy because the locus consisted of arcs with an angle of 90°. This is less than the 120° of the centre of the *treel*.

Step 3

At Gantug's clan's next harvest celebration they put burning flares on the corners of the *squeels*. What can you say about the loci and how their loci are related?

Give the students a chance to experiment again and use the rotation-by-rotation method (see Figure 2.22) to come up with their conclusions. Discuss what they find.

Figure 2.22: The locus of a flare on the corner of a rolling *squeel* using the rotation-by-rotation method

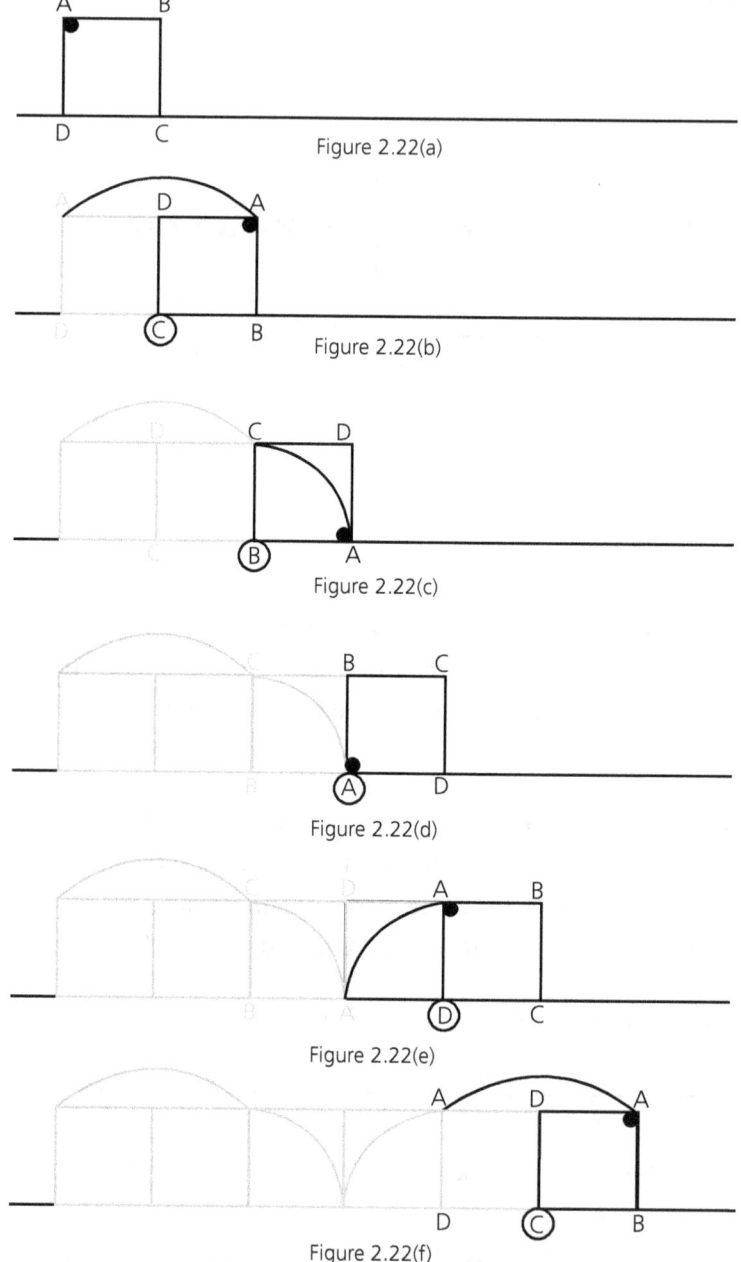

Figure 2.22(a)

Figure 2.22(b)

Figure 2.22(c)

Figure 2.22(d)

Figure 2.22(e)

Figure 2.22(f)

Note that the arcs look a bit like a double cloud that keeps getting repeated. Unlike the *treel* centre's locus, the *squeel* has arcs with various radii. Sometimes the arcs are quarter circles with a radius of one Gantug-arm, and sometimes they are quarter circles with radius bigger than one Gantug-arm (but less than two). The angles in the arcs are 90° because the *squeel* rotates through 90° every time it moves forward.

Step 4

But are the loci of each vertex the same?

Each locus is the same. It repeats after every four rotations, meaning each locus is a translation after one, two or three rotations of a chosen locus.

Step 5

Ask the students if they can tell you what the loci of the midpoints of a side of a *squeel* is. Get them to experiment and come up with their conclusions. The loci here are again all the same to within a translation. However, there are four distinct arcs (see Figure 2.23). What other differences are there from the loci of vertices of *squeels*? (Radii of arcs?)

Figure 2.23: The locus of a midpoint on the side of a *squeel*

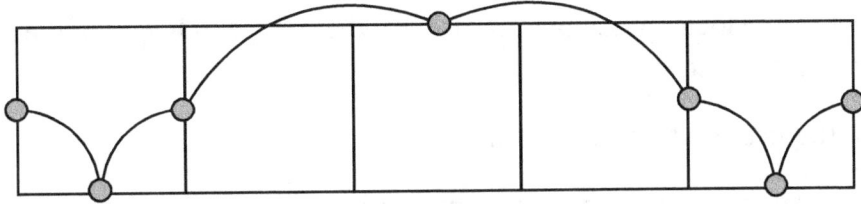

Step 6

Now let the class experiment with the loci of other points. What similarities and differences can they see between various interior points of the *squeel*? In Figure 2.24 we have given the locus of a point that is on the diagonal of the *squeel*.

Figure 2.24: The locus of an interior point that is never more than one side length above ground

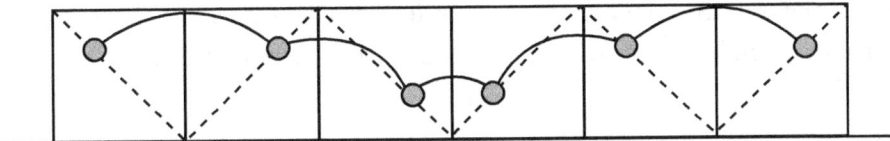

Where to from here?

- What happens to loci for points on *rectangular* wheels? (See the series website for a link.)
- What other problems can they invent?
- Which locus did they find most pretty and/or most interesting?

Level 4: Other wheels

Problem

It didn't take long for Gantug to realise that his invention would give a smoother and smoother ride if it had more and more sides. Regular pentagons were hard to make with the sort of circle that he had used with Fred, so he decided to make a wheel in the shape of a regular hexagon: a *heel*. Show that the *heel* does in fact give a smoother ride than a *squeel*.

Problem steps

Step 1

The locus drawing (Figure 2.25) shows that there are more arcs making up the locus but the distance that the locus gets above ground (i.e. the change in height of the locus) is less than the locus of the *squeel*.

Figure 2.25: The locus of the centre of a *heel*

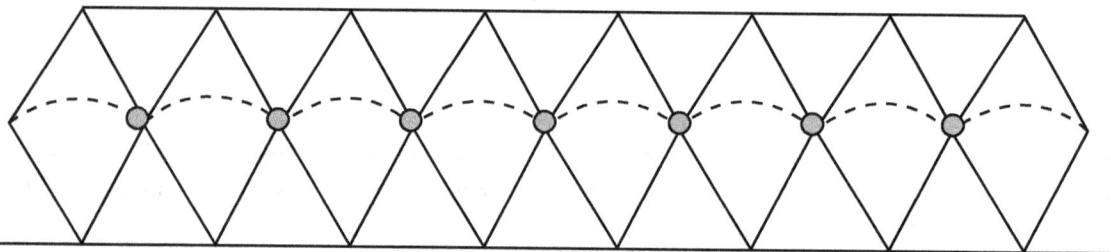

Once again the parts of the curve are arcs of circles with centres on the bottom horizontal line. The arcs have an angle of 60°, which is the secret of the better ride.

Step 2

But what about the vertices of the hexagon? What is its locus?

Figure 2.26: The locus of the vertex of a *heel*

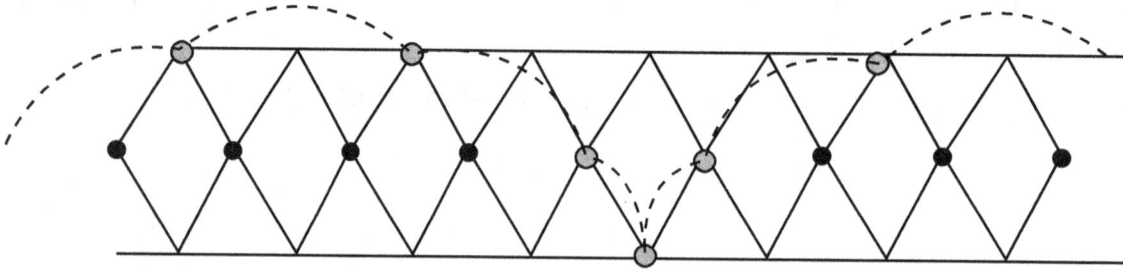

Figure 2.26 shows the locus of a vertex. This is a triple cloud. The arcs all have angle 60°, because that is the angle of rotation of the *heel* each time, but the radii of the arcs vary.

Step 3

Now experiment with the loci of different points on the *heel*. In what ways are they the same and what ways are they different?

Step 4

Finally, the penny dropped. One day, after reflecting on a hill near to the clan's latest site, Gantug decided that a *ceel* would be the best thing to have. A *ceel* is a wheel made out of a circle. Over time the name morphed into *wheel*.

Why is a *ceel* even better than a *treel* or a *squeel*? What does the locus of the centre of the *ceel* tell you? (It is a straight line.)

Experiment with loci on other points on the *ceel*.

Where to from here?

- Find the locus of any point on the circumference of the ceel. This is known as a cycloid. Finding the locus is not easy, but there are links to explanations and animations at the series website.
- What is the locus of a point inside the *ceel*? It is called a *curtate cycloid*. Notice how the curtate cycloid changes for the cycloid to the straight line.
- Can you find all of the sub-points of the *heel* and the *ceel*? (The *ceel* is easy.)
- What loci do you get by rotating a circle on the outside of a circle? (This is an *epicycloid*.) How about the inside? (This is a *hypocycloid*.) Check the series website for links with more information on these special cycloids.
- What other questions can your students think up?

CHAPTER 6:
BOXES AND MORE BOXES

Initial problem

How are cardboard boxes really made?

Background information

In this activity a lot of time is devoted to nets. At Level 1 these come in any guise; students investigate boxes that are used for commercial purposes and make boxes of their own. It can be done by Year 4 students.

From here we have a springboard to Level 2 into box construction via nets. This is a little harder but most Year 4 students and all students in Years 5 to 7 should be able to do it.

This is extended in Level 3 to discussing capacities of various boxes constructed from sheets of paper. Level 3 is still available for some Year 4 students but is primarily for Year 5 and above. It touches on some Year 8 concepts, but these should be accessible to students at Year 5 or above.

Finally, in Level 4 we investigate how we can get the biggest box out of a single piece of paper, aided by the use of spreadsheets. Level 4 is aimed at students in Years 6 and 7, but more able Year 5 students could also be involved if they know or can find out the volume of a cuboid.

Table 2.4: Australian Curriculum content descriptions for the *Boxes and more boxes* activity

Activity level	Problem	Content descriptions
1	Breaking and making boxes	*Year 4* Use scaled instruments to measure and compare lengths, masses, capacities and temperatures (ACMMG084) Compare the areas of regular and irregular shapes by informal means (ACMMG087) Compare and describe two-dimensional shapes that result from combining and splitting common shapes, with and without the use of digital technologies (ACMMG088) Create symmetrical patterns, pictures and shapes with and without digital technologies (ACMMG091) *Year 5* Connect three-dimensional objects with their nets and other two-dimensional representations (ACMMG111) Describe translations, reflections and rotations of two-dimensional shapes. Identify line and rotational symmetries (ACMMG114)
2	Boxes from rectangles	*Year 4* ACMMG084 (see above) ACMMG087 (see above) ACMMG088 (see above) ACMMG091 (see above) *Year 5* Choose appropriate units of measurement for length, area, volume, capacity and mass (ACMMG108) ACMMG114 (see above) *Year 6* Investigate, with and without digital technologies, angles on a straight line, angles at a point and vertically opposite angles. Use results to find unknown angles (ACMMG141) *Year 7* Calculate volumes of rectangular prisms (ACMMG160)
3	Boxes to size	*Year 4* ACMMG084 (see above) *Year 5* ACMMG114 (see above) *Year 6* Investigate combinations of translations, reflections and rotations, with and without the use of digital technologies (ACMMG142) *Year 7* ACMMG160 (see above) *Year 8* Develop the formulas for volumes of rectangular and triangular prisms and prisms in general. Use formulas to solve problems involving volume (ACMMG198)

»

Table 2.4: Australian Curriculum content descriptions for the *Boxes and more boxes* activity (continued)

Activity level	Problem	Content descriptions
4	The biggest boxes	*Year 6* Connect volume and capacity and their units of measurement (ACMMG138) ACMMG142 (see p. 85) *Year 7* ACMMG160 (see p. 85) Describe translations, reflections in an axis, and rotations of multiples of 90° on the Cartesian plane using coordinates. Identify line and rotational symmetries (ACMMG181) *Year 8* ACMMG198 (see p. 85)

Throughout this activity we use the pedantic word *capacity*, rather than the more commonly used *volume*. The reason for this is that the volume of a container is, strictly speaking, the amount of space that the container takes up. On the other hand, the capacity of a container is the amount of material the container will actually hold. So the capacity of a 1-litre container of milk tells us how much milk we can expect to be able to pour out of it and onto our breakfast cereal. The volume of the milk container itself is very small in comparison, and tells us nothing about whether we need to get another container of milk because we are expecting visitors for breakfast.

This distinction is not much observed in everyday language, where capacity is rarely used; but we use it because it is the correct technical term.

(There's a similar difficulty with the words weight and mass. Strictly speaking, mass is the amount of 'stuff' that a body has, measured in kilograms. This is the same wherever in the Universe the body happens to be at the time; you can wander off into space safe in the knowledge that your mass will be the same there as here. But weight is the amount of pull on your mass by the local massive body; pull is a force and is measured in Newtons. This pull changes depending on the local massive body. On Earth that body is the Earth—so if you really want to lose weight, as opposed to mass, your best strategy is to toddle off to the Moon.)

Big ideas

» Connect 3D objects with their nets and other 2D representations
» Explore 3D space

Suggested resources

» MAB or Base 10 blocks
» Black Line Master 1 cm grids

Problem aims

» To learn more about cuboids (rectangular prisms, boxes)

» To construct boxes for specific purposes from both given nets and nets developed by students

» To learn and/or apply formulae for the capacity of 3D objects

» To discover how to increase capacity while minimising the area of the constructing material

» To use a spreadsheet

Key concepts

» Square

» Rectangle

» Area and capacity

» Cuboid, rectangular prism

Possible heuristics/strategies

» Make a model

» Make a drawing

» Experiment to produce data

Level 1: Breaking and making boxes

Problem

How are cardboard boxes really made?

Problem steps

Prior to starting this activity, ask your students to bring along boxes that they have at home. Let them know that the boxes will be disposed of at school and won't be returned.

Step 1

This is a pure investigation. Ask your students to bring a wide collection of smallish boxes that have been made to hold things such as cereal, chocolates and other foods. Open them up in class to reveal the nets that have been designed to make them.

What do the manufacturers use to hold the boxes together? (Some use glue and some use tabs.) How does this affect their nets?

How easy is it to reproduce these boxes in the classroom? Get the class to construct a few boxes using the nets they have uncovered.

Step 2

In the next two steps, concentrate on boxes that are rectangular prisms (cuboids). You could also involve open (topless) boxes.

First, give the students an opportunity to make their own boxes from a net of their own design, using a minimum amount of glue to hold them together. Can they make boxes without using glue at all?

Step 3

Which boxes (open or closed) give the *largest* capacities for the *smallest* amount of cardboard?

Let the students theorise how this can be calculated. Should they scale the net size of each box to some standard size, then compare the capacities of the boxes? Might they do something seemingly strange, such as dividing the capacity of the box by the area of the paper that was cut to make it? Whatever criteria are used, the class should have a reason for using those criteria.

Step 4

Repeat steps 2 and 3 for boxes that are not cuboids/rectangular prisms.

Generally speaking, do the students seem to get more capacity from non-rectangular prisms?

Where to from here?

- Investigate the construction of pyramids. Is it better to have a base with a large number of sides? Do triangular pyramids give you the most capacity for your money?

Level 2: Boxes from a rectangle

Problem

Merinda makes a cuboid box that is open at opposite ends by folding a rectangle twice.

How does she do this? What is the capacity of such a box?

Problem steps

Step 1

Discuss with the class how Merinda can fold the rectangle to make a cuboid.

Figure 2.27: Folding a rectangle into a box

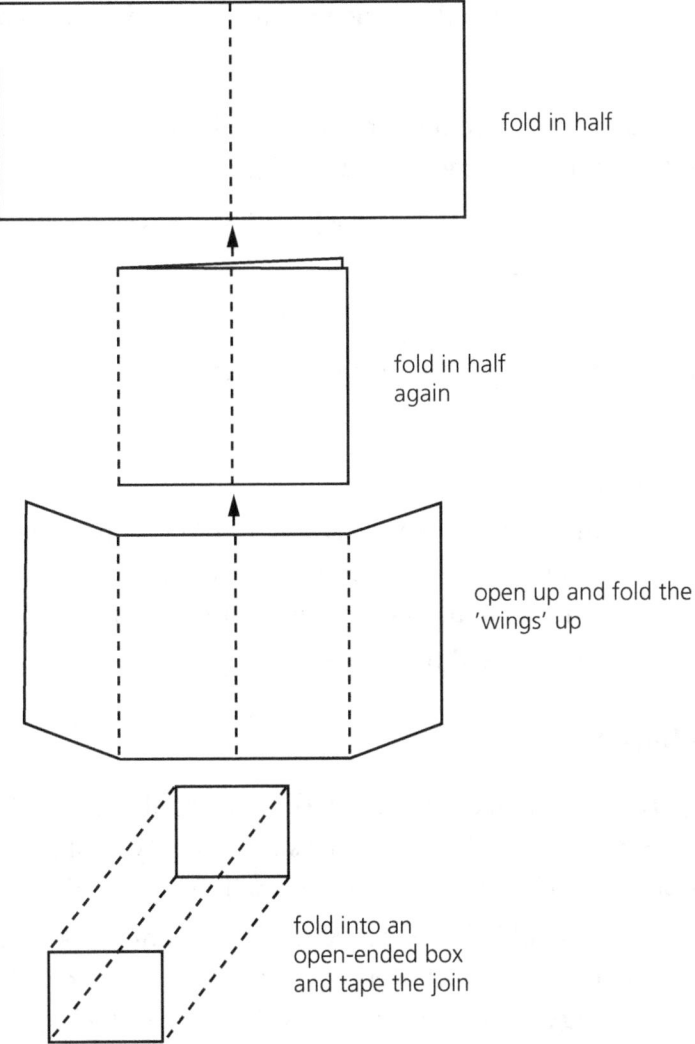

The method is shown in Figure 2.27. When the folded rectangle has been opened out, you need to refold one of the folds so that the box can be made. At this point, use tape to join the two edges of the rectangle together so that the box doesn't collapse.

Chapter 6: Boxes and more boxes

Step 2

Using the method in Step 1 (folding in half, then half again), get the students to make a box from a 4 cm x 8 cm piece of paper. Visit the series website and download the teacher file for this activity (1-cm grid paper). Print a sufficient number so that each student can construct their own box. How many different boxes can they make? (Two: 1 × 1 × 8 and 2 × 2 × 4.)

How many MAB or Base 10 mini blocks can fit into each box? (Eight for the 1 × 1 × 8 cuboid, 16 for the 2 × 2 × 4 one.)

Since the volume of one mini block is 1 cubic centimetre (cc), the capacities of the two open boxes are 8 cc and 16 cc, respectively.

Step 3

Repeat Step 2 starting from a 4 cm × 12 cm rectangle.

There will be two cuboids; one will be able to hold 12 minis and the other 36.

Why do two cuboids made from the same rectangle have different capacities? Discuss.

Step 4

Get the class to predict the capacities of the cuboids they would get from folding a 4 cm × 10 cm rectangle in the same way as in Step 2.

What if they fold other rectangles? Can they predict the capacities of the cuboids that come from a 100 by 200 rectangle?

Step 5

Discuss why you have been using the word *capacity* here, rather than *volume*.

Step 6

Put the students into pairs.

In each pair, one student suggests a capacity, such as 72 cc, and asks their partner what sized rectangle they would need to make a cuboid with that capacity. (Is there more than one answer to this?) Have the students swap roles within each pair.

You could extend this step by making this a quiz.

Where to from here?

| If students fold a rectangle twice they can make a triangular prism. Can they work out the capacity of this 'box'? Repeat for other prisms. What type of prism seems to give the biggest capacity? Which seems to give the smallest capacity?

| What other shapes can be folded like this? Can the students make different cuboids from the same sized grid paper as in steps 2, 3 and 4 (4 cm × 8 cm, 4 cm × 12 cm, 4 cm × 10 cm) by folding along the gridlines? Which gives the biggest capacity?

Level 3: Boxes to size

Problem

Merinda thought about the boxes that the class made from nets (see Level 1) and came up with this net. Here she cut out squares of the same size from a rectangle and then stuck each of the pairs of sides *A* and *B* together to make a box.

Figure 2.28: A net for an open box

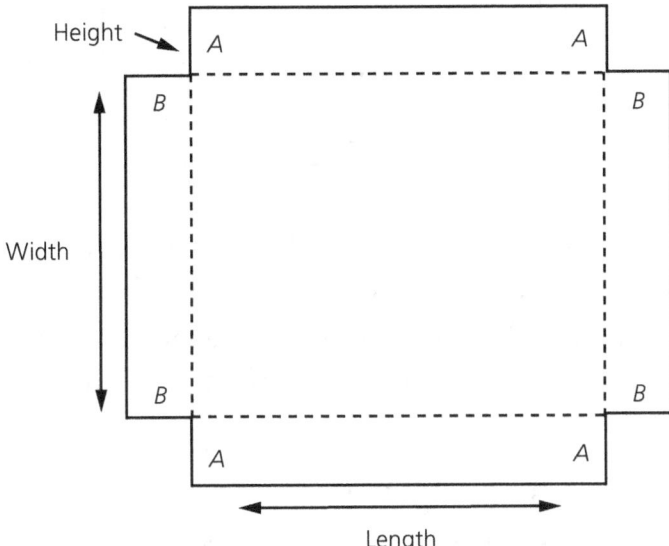

Can Merinda make a box to hold exactly 50 blocks? The blocks should be level with the top of the box when she is finished.

(Note: This question is related to 'Jenny's jellybean problem' from *Creative Activities in Mathematics Book 1*.)

Problem steps

Step 1

Use whatever blocks you have available as the basic unit of measurement. If you have 1 cm blocks (MAB or Base 10 blocks) and graph paper with 1 cm squares (the teacher file for this activity, available on the website series), that will make the process simpler and save a lot of calculating and measuring. That said, calculating and measuring are fundamental part of the curriculum, so don't eliminate them entirely from the activity.

One way for Merinda to build the open box is to have sides of 2 cm, 5 cm and 5 cm. We show the net that is needed in Figure 2.29. This can be made from a 9 cm × 9 cm square by cutting out 2 cm × 2 cm squares from each corner.

Figure 2.29: A net for an open box (with measurements)

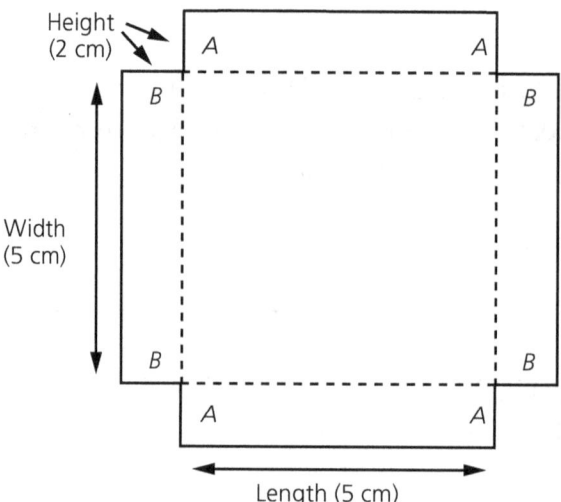

The rectangle of paper (cardboard) that Merinda needed to start with had dimensions 9 cm × 9 cm. The squares that she cut out were 2 cm × 2 cm.

Get the students to make the box as above with these measurements, then to test the box by filling it with cubes. They should find that the box holds exactly 50 cubes.

Step 2

Is this the only way to make a box for 50 cubes?

Let the class decide how they will make their own boxes for 50 cubes, either individually or in groups. When the first ones have finished, ask them whether there are any more ways to make a box for Merinda's cubes.

After everyone has made at least one box, get them together to discuss what they've done. How many different boxes can they think of? Make a list of these on the board, using a table like the one below. Discuss the results that your class has found.

Table 2.5: Possible boxes for Merinda's cubes

Length	Width	Height
1	1	50
1	50	1
1	2	25
1	25	2
2	25	1
1	5	10
1	10	5
5	1	10
5	10	1
2	5	5
5	2	5
5	5	2

(Note that a 1 × 2 × 25 open box is different from a 1 × 25 × 2 open box because of where the open part is. We can't make that distinction if the box is closed.)

All of the boxes in Table 2.5 can take exactly 50 of Merinda's cubes. So we say that they all have the same *capacity*—the amount that they can hold.

Get your class to design (but not necessarily make) open boxes that will take 120 small cubes. What is the capacity of such boxes? (120 cc or cm^3.)

Step 3

Now get students to draw a range of boxes with different whole-number dimensions.

Can they find the capacity of those boxes using MAB/Base 10 minis?

Can they find a quick way of finding the capacities? This is done by noticing that the number of cubes is the length × width × height.

Step 4

What happens if one of the sides is not a whole number? For example, what is the capacity of a box with length = $\frac{1}{2}$ cm, width = 1 cm and height = 1 cm?

It takes two of the $\frac{1}{2}$ × 1 × 1 boxes to make up a 1 × 1 × 1 cube. So the $\frac{1}{2}$ × 1 × 1 box must have capacity equal to $\frac{1}{2}$ of the 1 × 1 × 1 cube, or a capacity of $\frac{1}{2}$ cm.

What about a box of length = $4\frac{1}{2}$ cm, width = 2 cm and height = 3 cm?

Two of the $4\frac{1}{2}$ × 2 × 3 boxes can make a 9 × 2 × 3 box. We know that the capacity of a 9 × 2 × 3 box is 9 × 2 × 3 = 54. So the smaller box is half of that, or 27.

Discuss whether this fits the pattern of multiplying length by width by height.

Step 5

Is it *always* true that the capacity of a box can be found by multiplying together length, width and height? Discuss this with the whole class.

We don't prove this here but we can justify it for any fraction. This is a matter of stacking together enough of the fraction sides to make whole numbers, calculating the capacity of the combined boxes using the formula and then dividing by the number of small copies. This gives:

$$\text{Capacity of a box} = \text{length by width by height}$$

This can be formalised. Things get more difficult if we have a side that is not a fraction—for instance, a side with length π—but even then the formula holds.

Where to from here?

- Is there any difference between the capacity of an open box and the capacity of a closed box with the same dimensions?
- What is the volume of a mattress? What other capacities or volumes can the class now calculate? Can they find the capacities of shapes other than boxes (rectangular prisms)?

Level 4: The biggest boxes

Problem

Merinda has a square piece of paper with side length 10 cm. What is the biggest capacity box she could get from that piece of paper by cutting small squares from the corners?

Problem steps

Step 1

Try out a few values and see what Merinda gets for the capacity. If she cut off a 1 cm × 1 cm square from each corner she would get a box that is 8 × 8 × 1 cm.

What if she cut off some bigger squares? Does the capacity always increase? (Don't let the students do all possible squares.)

Step 2

Ask your students what calculations they did to alter the dimensions of the box.

Try some different-sized squares. Then ask them to take off a square with side length x and calculate the dimensions. (The dimensions of the box here are $10 - 2x \times 10 - 2x \times x$.)

Step 3

What is the biggest capacity they can find?

This is a good time to incorporate Excel/spreadsheets into the activity and save time doing calculations.

1. Open up an Excel sheet. In cell A1 put the heading 'Height x', in cell B1 'Length 10 – 2x', in C1 'Width 10 – 2x' and in D1 'Capacity of the box'.
2. Then drag down from A2 to give the x entries 1, 2, 3, 4, 5. There is no need to go further than this. (If $x = 5$, then Merinda will have no box.)
3. In the B and C columns drag down to enter 8, 6, 4, 2, 0 to correspond with the entries in column A.
4. We know that the capacity of the box is $x \times 10 - 2x \times 10 - 2x$, so enter '=A2*B2*C2' in cell D2. If you drag down from D2 the entries will be the capacities of the various possible boxes.
5. This will give a biggest capacity of 72 when $x = 2$.

Step 4

Your students can increase their spreadsheet's accuracy by using decimals for the height instead of whole numbers.

To save time, they only have to look at numbers around 2. Ask them to replace the x entries in column A with 1.3 to 2.4. When the spreadsheet makes the calculation, it shows that $x = 1.7$ produces the maximum capacity.

If you want them to be even more accurate, use 1.59 to 1.72 as the values for x, giving a maximum capacity for $x = 1.67$.

More and more accuracy will suggest that the value of x that gives maximum volume (about 74.070) is 1.6667 or $1\frac{2}{3}$.

(The great value of spreadsheets is that once you have set up a scenario, it is easy to make slight changes and do lots of related problems. These may help your students make predictions about more general problems.)

Where to from here?

- Step 4 suggests an investigation. If a 10 × 10 piece of paper can produce a maximum value of about 74 when $x = 1\frac{2}{3}$, then there might be a pattern with square paper. What capacity would you expect from a 100 × 100 piece of paper, or an $n \times n$ piece? (With an $n \times n$ square we get a maximum capacity of $\frac{2n^3}{27}$ when $x = \frac{n}{6}$.)

- The spreadsheet can be adjusted to make predictions about rectangles too. Let them try out a few and see what they get.

- This investigation can be extended to rectangles with sides m and n. If the squares are cut off in the same way and the box made by folding, then the maximum volume is found when the side of the square is $\frac{(m + n) - \sqrt{m^2 - mn + n^2}}{6}$. (This is difficult to obtain by guessing even from a large number of examples.)

- Can your students now see other ways to use spreadsheets?

PART 3: STATISTICS AND PROBABILITY

Part 3 presents three activities centred on the Statistics and Probability strand.

Table 3.1: Statistics and Probability activities

Problem	Big ideas
Sports shots	▪ Increasing the number of trials will also increase the accuracy of the predicted probability
Roman gamblers	▪ Conjecturing and justifying
	▪ Being systematic
	▪ Fairness
The school fete	▪ Collecting data and displaying it to make conclusions

Some reminders before you use these tasks in your classroom:

1. The questions in the text are ones you can ask your students. You're likely to be able to produce similar, more immediately relevant ones for your particular students as you work on these activities with them.
2. We have given suggested links to the Years in the Australian Curriculum: Mathematics for all the Levels in each activity. But given that there will be a spread of ability in your classes you should take these as a guide only. Take the opportunity to encourage every student to the edge of their comfort zone.
3. To take all students further, sometimes you can omit some of the later steps of a Level in favour of the early steps in the following Level.

CHAPTER 7: SPORTS SHOTS

Initial problem

How far can your students throw a ball?

What are the chances of them throwing it 10 metres, 20 metres, 30 metres or more than 30 metres?

How are these events related?

Background information

This activity gives students a chance to combine physical activity and mathematics, letting them engage in various sporting activities and using the measurements taken as a basis for predictions. This activity is a crossover between both physical education and mathematics, incorporating measurement and probability.

In Level 1, students try to throw balls for set distances, then throw balls to see what range of distances they can throw, and determine probabilities from the results. Level 1 can be done by Year 4 students and up.

The mathematics of Level 1 is repeated at Level 2 with the introduction of fractions to describe probabilities. Much of Level 2 can be done by many Year 4 students, while all Year 5 students should be able to complete the whole activity.

For Level 3 the work extends to writing probabilities using fractions, decimals and percentages. Here the data is produced using four stations, with whatever physical activities are available. Level 3 is appropriate for Year 5 students.

Finally, in Level 4, the focus shifts to accuracy by looking at kicking for goal and then creating a fair game as a result. Level 4 can be done by some Year 5 students and all students in Years 6 and 7.

Table 3.2: Australian Curriculum content descriptions for the *Sports shots* activity

Activity level	Problem	Content descriptions
1	Throwing	*Year 4* Describe possible everyday events and order their chances of occurring (ACMSP092) Identify everyday events where one cannot happen if the other happens (ACMSP093) Identify events where the chance of one will not be affected by the occurrence of the other (ACMSP094) Use scaled instruments to measure and compare lengths, masses, capacities and temperatures (ACMMG084)
2	Kicking	*Year 4* ACMMG084 (see above) *Year 5* List outcomes of chance experiments involving equally likely outcomes and represent probabilities of those outcomes using fractions (ACMSP116) Recognise that probabilities range from 0 to 1 (ACMSP117) Pose questions and collect categorical or numerical data by observation or survey (ACMSP118) Describe and interpret different data sets in context (ACMSP120)
3	Quantifying and comparing	*Year 6* Describe probabilities using fractions, decimals and percentages (ACMSP144) Conduct chance experiments with both small and large numbers of trials using appropriate digital technologies (ACMSP145)
4	Set shots for goal	*Year 6* ACMSP144 (see above) *Year 7* Assign probabilities to the outcomes of events and determine probabilities for events (ACMSP168)

It may be more efficient to collect the data for Levels 1 and 2 at the same time. The Level 2 data might be used later in the year.

We have put the emphasis on football games, especially in Level 4, but there is no reason why hockey or any other game can't be the basis for the work.

If some of your students are unable to be involved with the physical aspects of the activity, or may be embarrassed by being involved, you might want to restrict the involvement to volunteers only. The remaining students would then be involved in collecting and recording the data.

(However, in our experience the majority of students will be more than happy to engage in sport during their mathematics lessons!)

Big ideas

» Collecting data

» Assigning numbers to various probability categories

Suggested resources

» Workbook
» Calculators
» Measuring tapes (ideally ones that can measure more than 10 metres)
» T-balls/footballs/soccer/basket/net balls
» Witches' hats

Problem aim

» To create an authentic, real-world opportunity for students to engage in probability and statistics

Key concepts

» Collecting and interpreting data
» Understanding that increasing the number of trials will also increase the accuracy of the predicted probability

Possible heuristics/strategies

» Record your work
» Use a table or graph
» Act it out
» Use equipment
» Be systematic

Level 1: Throwing

Problem

How far can your students throw a ball?

What are the chances of them throwing it 10 metres, 20 metres, 30 metres or more than 30 metres?

How are these events related?

Problem steps

Step 1

Before leaving the classroom, give the students time to think about what is going to happen, what data they are going to collect, how they will collect it and what the data will tell them. Impress on them that they are trying to throw a ball as far as possible in a given direction and that the distance is calculated from the throw line to where the ball first hits the ground. Is it likely that any student will be able to throw the ball the full 10 metres, 20 metres, 30 metres or more than 30 metres?

How will students collect this data? Should they pool the data for the whole class or keep each student's data separate? What are they going to be able to conclude if they pool the data or if they don't? Should they exclude very short throws or very long ones? How will they quantify their chances of success—that is, the chance of getting into each category? (They will need to compare the number of times they got the ball in a certain range to the number of throws they each had.) The students should discuss these points and produce reasons for their decisions.

Also get them to think about the *reliability* and *practicability* of the data they collect. Should they only have one throw each or 10 000 throws? These should also be discussed and reasons provided for their conclusions. It is important for students to realise that people rarely throw a ball—or do anything else—in exactly the same way every time, so there is going to be variation in their throws. This is the foundation of statistics and probability, and also of life. If the same sports teams always performed the same way, there would be no reason to watch the games they played. Probability and statistics allow us to see how things vary and to try to predict the chances of certain things happening.

Let each student write down what they think their chances are of throwing the ball these distances. (It might be useful for them to describe how far they think 10 metres, 20 metres and 30 metres are before they do this. For example, a classroom is 10 metres long and the distance from one end of the playground to the other is 30 metres. These might be checked before they start throwing.) These chances should be categorised as:

- unlikely to happen
- least likely to happen
- most likely to happen
- certain to happen.

Once the students have made predictions and discussed concepts, move outside to the play area and divide the class into groups. Each group's first task is to mark a throwing line and then lines 10, 20 and 30 metres away from this. (It is worth the class double-checking these measurements and comparing the measurements of different groups to make sure that there are no large discrepancies.)

Each student then takes turns in both throwing the ball the number of times that has been decided beforehand and collecting the data.

Record the number of times the throws are in each of the following distance categories:

(a) the ball didn't go 10 metres

(b) the ball went between 10 metres and 20 metres

(c) the ball went between 20 metres and 30 metres

(d) the ball went further than 30 metres.

Step 2

On returning to the classroom, students in each group should use the recorded data to place each event in one of the four distance categories of Step 1. How could they display this data? This could be done in a table or using a dot graph. We have shown examples of these for 30 throws in Figure 3.1 below.

Table 3.3: Representing throwing data in a table

Number of throws	(a)	(b)	(c)	(d)
Angelo	4	14	10	2
Bella	8	20	2	0

Figure 3.1: Representing throwing data in a dot graph (Angelo recorded as diamonds; Bella as dots)

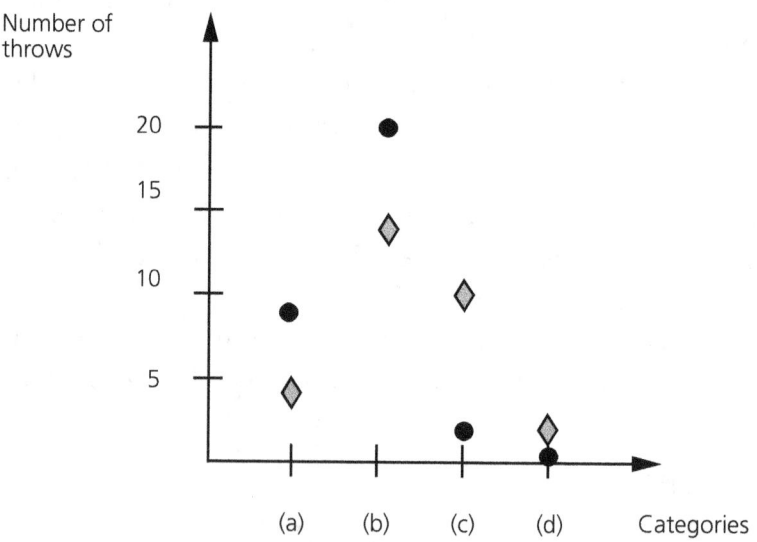

What conclusions can the class make from these results? (Event (b) is more likely than the other events; most throws are between 10 and 20 metres etc. More specifically, from the data in Figure 3.1, (b) is the most likely to happen and (d) is the least likely.)

The class can also compare events. For example, let them investigate the chances of different things happening. From Figure 3.1 they should see that:

- event (d) is less likely to happen than (a)
- event (a) is less likely to happen than (c)
- event (c) is less likely to happen than (b).

Can they produce events that are unlikely to happen or certain to happen?

Step 3

What conclusions can be made when all of the class data is combined? You should be careful to say under what circumstances they made their conclusions. For example, are the conclusions true for *all* children of your class's age? Are they only true for your class? Would the distances change if different balls were used or if the weather had been different?

What other situations can the class think of that are *dependent* on specific circumstances? (For example, whether a 12-year-old plays AFL or NRL may depend on where they were brought up; the colour of somebody's eyes will depend on the colours of their parents' eyes.) Let them work on this in groups and report back.

Variables that depend on other variables are called *dependent variables*. So the variable 'distance you can throw a ball' depends on the variable 'what kind of ball you throw'. Let them go back to the examples they found above and say which variables are dependent.

On the other hand, some variables are completely independent of others. These are called *independent variables*. Get groups to explore what independent variables they know. (For example, events (a), (b), (c) and (d) on page 102; the chance of Angelo throwing between 25 and 40 metres is not affected by Bella throwing just 25 metres; rolling a 1 or a 6 with a die.)

Step 4

From the data, can your students now find events that are:

- unlikely to happen
- least likely to happen
- most likely to happen
- certain to happen.

For example, it is certain that the ball will be thrown at least 10 metres, it is most likely that the ball will be thrown between 10 and 40 metres, and it is certain that some students can throw the ball further than you, the teacher.

What events that have nothing to do with the throwing exercise could fit into these categories?

Where to from here?

- Students might like to compare the distances that they can throw different balls, such as a T-ball, a tennis ball or a table-tennis ball. Which can they throw furthest and by how much? Can they explain these various lengths?

- How does the weight and size of a ball affect the distance you they throw it? What other things affect the distance the ball can be thrown? (Shape of ball, height of thrower, angle of projection, flatness of the land, strength and direction of wind?) There is an opportunity here to bring scientific aspects into the discussion.

- Ask the class which of the data representations they found most useful and why.

- It might be worth looking at the advantages of training on throws. For example, if students are given a chance to practise throwing during physical education classes, will they improve the distance they can throw? If so, by how much? This could be done experimentally. The same thing might be done for learning a new skill, such as serving the ball in tennis.

Level 2: Kicking

Problem

Divide your students up into groups and let them go outside to see how far they can kick a ball.

How far might they be able to kick it? What are the chances of them kicking various set distances?

('Distance' here is the distance until the ball hits the ground for the first time.)

Problem steps

Step 1

Before leaving the classroom, give the students time to think about what might happen, what data they might collect, how they will collect it and what the data will tell them. (In many ways this is a repeat of Level 1, Step 1.) Base the data collection again on whether the ball went:

(a) less than 10 metres

(b) between 10 metres and 20 metres

(c) between 20 metres and 30 metres

(d) further than 30 metres.

How will they collect this data? (This should be much as it was collected in Level 1, Step 1.)

Also get them to think about the reliability and practicability of their data collection, and about independent and dependent variables as well.

We usually use numbers to tell us the probabilities of certain events. If an event is extremely unlikely to happen we say it has 'no chance', so logically we give that a value of zero. And if something is certain to happen it will happen 10 times out of 10; $\frac{10}{10} = 1$, so certain events are given the number 1.

Discuss with your class what number they would give to something being equally likely. What is the chance of a baby being a girl or a boy? On average, 5 babies out of 10 will be female, so we give $\frac{5}{10} = \frac{1}{2}$ to equally likely chances.

Follow this up with fractions for 'not very likely to happen' and 'most likely to happen, but not certain'.

Now turn things around. What words would students associate with the probabilities $\frac{1}{10}, \frac{2}{10}, \frac{3}{10}, \frac{4}{10}, \frac{5}{10}, \frac{6}{10}, \frac{7}{10}, \frac{8}{10}$ and $\frac{9}{10}$?

Using these categories, let each student write down what they think their chances are of kicking the ball the distances shown in (a), (b), (c) and (d) above.

Once the students have made their predictions, move outside and divide the class into groups. Give everyone the opportunity to kick the ball 12 times and keep a record of their ten best kicks.

Step 2

On returning to the classroom, let each group use the recorded data to produce a probability *in terms of fractions* for each of the four events on page 105. This should be done for individuals and for the group as a whole.

Combine these probabilities for the class as a whole, and then compare group probabilities with the class probabilities. If one group has a probability of $\frac{28}{40}$ for kicking a ball between 10 and 20 metres, and the class as a whole has a probability of $\frac{41}{280}$, is that group better than the class at that distance or is it the other way around?

Step 3

What fractions point to outcomes being unlikely to happen, not very likely to happen, equally likely to happen, most likely to happen or certain to happen? What events were unlikely to happen, not very likely to happen, equally likely to happen, most likely to happen or certain to happen? Does this differ in any way from the fractions you used before the kicking took place? Why have they changed (or not changed) your opinion?

What fractions were used in making the probabilities? What was the highest fraction used? What was the lowest fraction used? Did you ever use something like $\frac{25}{20}$? Why not?

(The probabilities were made by dividing the number of successful kicks for that distance by the total number of kicks; so the fraction cannot be bigger than one.)

Step 4

By now students should have worked out the probabilities for each of the four distance categories from Step 1. What is the sum of these four probabilities for individuals and for the class as a whole?

Let the students do the calculations. They may make errors here, but they should all reach a sum equal to 1.

Why is this answer 1? What does a probability of 1 mean?

(As there were a certain number of kicks, and each of these kicks was recorded in one of the four categories, the sum of the probabilities is the total number of kicks divided by the total number of kicks. This has to be 1.)

Step 5

Is it true that all boys can kick further than all girls? Discuss how you would decide this.

What about left-footed kickers and right-footed kickers?

Where to from here?

- What conclusions can students come to here?
- What differences are there between the distances kicked by the balls used in the various codes of football? (There are various scientific explanations for this that could be considered.)

- If the students measure the *actual* distance that the balls have gone, what shape graph might they get? Display this data by putting the actual distance, to the nearest five metres, along the horizontal axis as shown in Figure 3.2. Then put the number of times that distance was kicked on the vertical axis.

Figure 3.2: An example dot graph of distance versus number of kicks

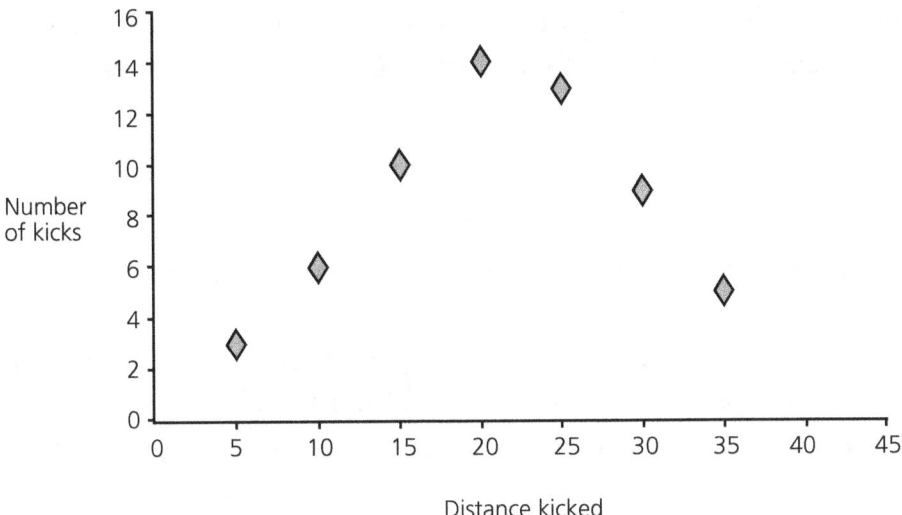

- Is there an explanation for this? What has it got to do with the *normal* (bell, Gaussian) curve? (The reason is that most people can do things in a certain middle range. Very good kickers can kick the ball further than average, but there aren't many good kickers, so large distances are kicked less often. Note that many things follow a normal curve—height, weight, yearly earnings, test marks etc.)

Level 3: Quantifying and comparing

Problem

Set up three or four sports stations outside. These may involve throwing, hitting, kicking or anything else that you prefer (or that is possible using school equipment).

Decide on what data you will collect and what events you will determine probabilities for. What probabilities would your class expect, in terms of fractions, decimals and percentages, for each event? How good are they at predicting these probabilities?

Problem steps

Step 1

We assume that you can set up the following stations, but feel free to work within whatever constraints you have.

- **Station one, cricket:** Set up three sets of stumps. Let each student throw a cricket (or similar) ball, either under- or over-arm, at the stumps from 10 metres in front.
- **Station two, hockey:** Set up a hockey goal. Let students hit a hockey ball from 15 metres at the goal.
- **Station three, netball:** Shoot for goals into netball rings. A score is recorded if the ball scores a goal or if it is a near miss.
- **Station four, tennis:** Place a table or other obstacle, about 2 metres long and 2.5 metres tall, 15 metres away from the hitter. Students hit a tennis ball over the obstacle, first bouncing the ball then using an underarm stroke. They score a point if the ball goes over the obstacle.

The students need to decide what events (in the probability sense) they want to concentrate on. Their aim is to see what probabilities of success they have at each station—but what does this mean? Have a class discussion before they use the equipment so that they understand what they are trying to do and how much data they want (need) to collect. These probabilities should be recorded in fractions, decimals and percentages.

Before going outside, let them all guess what their own probabilities will be and what will be the probabilities of the group and the class as a whole.

Step 2

Assign students into groups and send them to the stations to collect and analyse the data.

They should also check for errors in their calculations. One simple check is that all possible events should always sum to 1, 1.00 or 100% (depending on whether fractions, decimals or percentages are used).

Step 3

Get each group to present their data and conclusions to the class using appropriate visual representations. The representations may be any of the ones used above or any that you may have introduced in class. You might get the class to discuss which representation they find easiest to understand.

Step 4

Put the results on the board near to the predictions. How close are the actual probabilities and the predicted probabilities? Why?

How could the students improve their predictions in future? Could they improve their data collection? How do they think they would do this?

Step 5

Predict the probabilities for another class at your year level. Get them to collect data and then see how good your predictions were. How can the predictions be improved?

Where to from here?

- What conclusions can your class make from the experiments they have made for this level?
- Which part of this work do they think was the hardest? Which part did they find easiest to understand?

Level 4: Set shots for goal

Problem

What is the probability of you kicking a goal from a set shot?

(A 'set shot' is a shot without interference from a specified point on the field.)

Problem steps

Step 1

Take the class outside to the school football field.

The whole class should work together to position four witches' hats in positions around the goalposts. They need to place each hat in a position that creates a varying level of difficulty for each set shot. They mustn't place the hats in positions that make the probability of kicking a goal either certain or impossible.

Step 2

Once the position of the set shots has been established, each member of the class will take ten set shots from each of the different positions.

Before taking the shots, students must decide on the best method of recording the data they collect. They must include the experimental probability for each of the set shots in fractions, percentages and decimals.

Step 3

Once this information has been recorded, have students answer the following questions:

1. Based on the data collected, who has the greatest probability of kicking a goal from each of the different locations?
2. How could they alter the experiment to make the probability more reliable?

Step 4

Place the students into small groups.

Based on the data collected, get each group to create a reasonably fair game involving kicking for goal from set shots. The group must explain their game and write a set of rules with reference to the data that was collected by the whole class.

Step 5

Each group trials their game and records the results.

Do the results reflect their expectations? If so, they should show mathematically how the results of the game indicate its fairness. If not, alter the game to make it fair.

Step 6

You could then try a tennis game, where each player had a given probability of winning on their serve and then another probability of winning when the ball is in play.

After that you might extend the game back to football, with players in positions around a football field. When a player gets the ball she has an 80% chance of passing to the nearest player and a 20% chance of passing to an opposition player. Nearer the goal, similar percentages apply for kicking a goal or a point.

Where to from here?

- What probability sums of your data add to 1?
- What was the strangest fraction that you had for a probability? Why was it strange?
- What events did you think you needed to collect more data for? Why?
- Students might investigate AFL, soccer or netball players online and research the probability of certain players scoring goals from various locations. They could then rank the players in terms of their scoring prowess, or draw other inferences from the data.

CHAPTER 8: ROMAN GAMBLERS

Initial problem

Augustus and Julius, two Roman soldiers, had a regular gambling session when they were on a campaign. It relieved the boredom of marching all day and subduing the local peasants. They and their compatriots gambled till very late every night.

But Julius was getting annoyed with Augustus, who he suspected was cheating. So one night around the camp fire, Julius made another die; he took his knife and quickly cut a cube from a small branch, marking the six sides with numbers.

What did the die that Julius made look like?

Background information

This problem is about developing the ideas of conjecturing and proving, as well as looking at simple probabilities with usual and unusual dice. The use of symmetry and being systematic are fundamental to this activity. One specific goal of this activity is to count possible dice that occur in different situations; another is to find the probabilities of the different possible events when the dice are rolled. This counting will lead on to theoretical probability later.

In Level 1, the soldiers use only the numbers 1 and 2 on their dice. This Level may be done by some students in Year 5, but is more accessible to older students who have learned the techniques involved.

In Level 2 the soldiers use 1, 2 and 3 on the dice. Again, this is more suited to Year 5 and 6 students.

In Level 3 the dice use the numbers 1 to 6, restricted in how they can be placed on the die so that opposite sides add to 7. Level 4 also uses 1 to 6, but the numbers are not restricted. Both of these Levels are for students in Years 6 and 7.

Table 3.4: Australian Curriculum content descriptions for the *Roman gamblers* activity

Activity level	Problem	Content descriptions
1	Simple Roman dice	*Year 5* List outcomes of chance experiments involving equally likely outcomes and represent probabilities of those outcomes using fractions (ACMSP116) Recognise that probabilities range from 0 to 1 (ACMSP117) Pose questions and collect categorical or numerical data by observation or survey (ACMSP118) Connect three-dimensional objects with their nets and other two-dimensional representations (ACMMG111) *Year 6* Describe probabilities using fractions, decimals and percentages (ACMSP144)
2	More Roman dice	*Year 5* ACMSP116 (see above) ACMSP117 (see above) ACMSP118 (see above) *Year 6* ACMSP144 (see above) Investigate combinations of translations, reflections and rotations, with and without the use of digital technologies (ACMMG142)
3	How many proper dice?	*Year 6* ACMSP144 (see above) ACMMG142 (see above) *Year 7* Assign probabilities to the outcomes of events and determine probabilities for events (ACMSP168) Represent events in two-way tables and Venn diagrams and solve related problems (ACMSP292)
4	How many wonky dice?	*Year 7* ACMSP168 (see above) Describe translations, reflections in an axis, and rotations of multiples of 90° on the Cartesian plane using coordinates. Identify line and rotational symmetries (ACMMG181)

Dice have been around for a very long time; the earliest known die is 5000 or so years old. Gaming with dice is known to have existed in many ancient cultures and the Romans especially were ardent gamblers. More information about the history of dice can be found online, and might be a good topic for a project, potentially looking at all solids that have been used to make dice. (Students interested in role-playing games might also be familiar with polyhedral dice.)

Big ideas

» Conjecturing and justifying
» Being systematic
» Using symmetry
» Simple probability

Suggested resources

» Blank cube shapes
» Printable cube and net diagrams (downloadable from the series website)

Problem aims

» To develop conjecturing and proof
» To see how to use symmetry
» To learn to be symmetric
» To learn how to count unusual objects

Key concepts

» Conjecturing
» Proving
» Posing questions

Possible heuristics/strategies

» Be systematic
» Use drawings
» Use materials

Special notes

Conjecture: A conjecture or guess is an attempt to find what might happen in a given situation. The 'situation' here is to find how many ways dice of certain kinds can be made. The conjecture for a specific situation might be 5. This might have been suggested by using concrete materials to get an idea of the number involved, but some possibility may have been overlooked when making the concrete materials. So until a proof is found, using some theoretical method, it is not possible to be sure that the conjecture is true.

Justification: Justification and proof are very similar but we don't expect a solid, formal proof. In this activity we use two techniques for justification—*being systematic* and using the *symmetry* of a cube. By looking at everything that might happen (being systematic), we sort out only the ones that *can* happen. Symmetry is then used to make sure that we haven't counted a given die twice.

Level 1: Simple Roman dice

Problem

Augustus and Julius, two Roman soldiers, had a regular gambling session when they were on a campaign. It relieved the boredom of marching all day and subduing the local peasants. They and their compatriots gambled till very late every night.

But Julius was getting annoyed with Augustus, who he suspected was cheating. So one night around the campfire, Julius made another die; he took his knife and quickly cut a cube from a small branch, marking the six sides with numbers.

What did the die that Julius made look like?

Problem steps

Step 1

Place students into small groups and distribute the net diagrams (printable versions are downloadable from the series website) to each group. Students should cut out the nets, write numbers on the faces and then glue the nets into cubes.

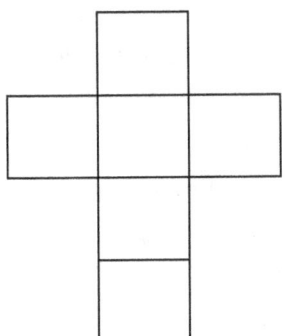

The point of this problem is to find the properties of the six-sided dice that are used in board games such as *Snakes and Ladders*, *Monopoly* and so on. You may want to talk about those dice, or even distribute some as examples.

The key things to come out of the ensuing classroom discussion should include those listed below.

- Most dice are cubes, though they can be other shapes.
- Each face of the cube has a number on it.
- Each of the numbers 1, 2, 3, 4, 5 and 6 is used.
- The numbers on opposite faces of a 'proper' die add to 7.

Step 2

Julius decided to do something different to break his losing streak. On one face of the die he marked the number 1, while on the other five faces he put the number 2. How many different dice could Julius have made?

In their groups, students should make as many dice as they can and conjecture the number of possible dice that Julius could make. Each group then discusses their conjectures with the whole class, justifying these conjectures using their models or otherwise.

There is in fact only one such die. Take any two dice made or presented by students and rotate them until the face with the 1 on it is on top. Both dice will look like the one in Figure 3.3. This is essentially due to the symmetry of the cube.

Figure 3.3: Julius' die with one 1 and five 2s

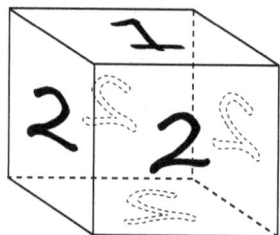

You might point out that this die also differs from the standard dice discussed in Step 1, in that opposite faces of the die don't add up to the same number.

Step 3

What is the probability of getting a 1 or a 2 with Julius' die? (For a 1: $\frac{1}{6}$; for a 2: $\frac{5}{6}$.)

If Julius made two dice of this type, what sums would he get when he rolled them? What probability would each of these sums be? (Possible sums are 2, 3 and 4. The probabilities are for a 2: $\frac{1}{36}$; for a 3: $\frac{10}{36}$; for a 4: $\frac{25}{36}$.)

What is the sum of *all* of the probabilities, both in the one-die and two-dice cases? (The totals of $\frac{1}{6}$ and $\frac{5}{6}$ are 1, as are the sums of $\frac{1}{36}$, $\frac{10}{36}$ and $\frac{25}{36}$.) Encourage your students to use this as a check for getting the probabilities correct in later steps.

One quick way to find the probabilities is to use a table like Table 3.5.

Table 3.5: Outcomes of rolling two of Julius' dice

	1	2	2	2	2	2
1	2	3	3	3	3	3
2	3	4	4	4	4	4
2	3	4	4	4	4	4
2	3	4	4	4	4	4
2	3	4	4	4	4	4
2	3	4	4	4	4	4

Step 4

Augustus *had* been cheating and very much wanted to keep playing with his own die, so he strongly objected to Julius' die. Julius responded by making a new die with two 1s and four 2s. How many dice could he have made?

As in Step 2, let student groups experiment by making dice with this set of numbers. Get them to conjecture how many there are and justify their conjectures (models or otherwise).

There are two possible dice here, as shown in Figure 3.4. Either the ones are on opposite faces or on adjacent faces with a common edge. Then any die with two 1s and four 2s can be rotated to look like one of the dice below.

Figure 3.4: The dice with two 1s and four 2s

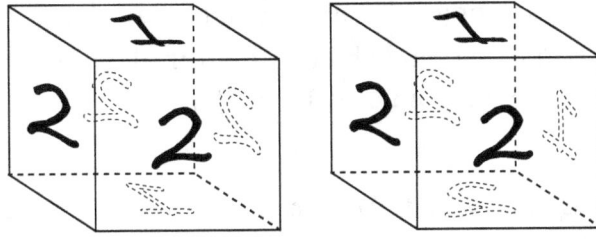

Step 5

What is the probability of getting a 1 or a 2 with each of Julius' dice? (They give the same probabilities—for a 1: $\frac{2}{6}$; for a 2: $\frac{4}{6}$.)

If Julius made two dice like either one in Figure 3.4, what sums would he get when he rolled them? What probability would each of these sums be? (For each die the sums are 2, 3 and 4. The probabilities are for a 2: $\frac{4}{36}$; for a 3: $\frac{16}{36}$; and for a 4: $\frac{16}{36}$.)

What is the sum of *all* of the probabilities, both in the one-die and two-dice cases? (The total of $\frac{2}{6}$ and $\frac{4}{6}$ is 1, as is the sum of $\frac{4}{36}$, $\frac{16}{36}$ and $\frac{16}{36}$.)

Step 6

Augustus was fuming by this point, but Julius was having fun, so he ignored Augustus and tried another tactic. 'Maybe I'll get a die you'll be happy with if I use three 1s and three 2s.'

How many dice are there with three 1s and three 2s?

Once again, let groups experiment by making dice with this set of numbers, conjecture how many there are and justify their conjectures.

There are two possible dice again, as shown in Figure 3.5.

Figure 3.5: The dice with three 1s and three 2s

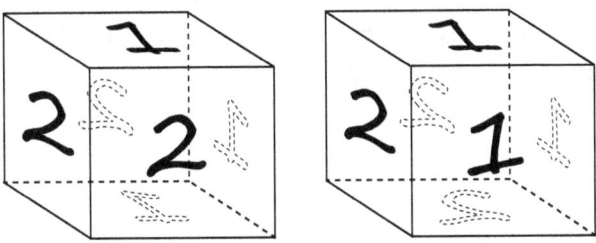

This is a little harder to justify. First, think about the 1s. Two of these are either on opposite faces or adjacent faces. If they are on opposite faces, then where can the third one be? Wherever it is, the die can be rotated until the ones are in the first position in Figure 3.5. Then the 2s have to be on the other faces and we get the die in the first position of Figure 3.5.

If no pair of 1s is opposite each other, then the faces they are on have a common vertex. The 2s can then be entered to give the other die in Figure 3.5.

Step 7

What is the probability of getting a 1 or a 2 with each of the dice in Figure 3.5? (For a 1: $\frac{3}{6}$; for a 2: $\frac{3}{6}$.)

If Julius made two dice like either one in Figure 3.5, what sums would he get when he rolled them? What probability would each of these sums be? (For each die the sums are 2, 3 and 4. The probabilities are for a 2: $\frac{1}{4}$, for a 3: $\frac{1}{2}$ and for a 4: $\frac{1}{4}$.)

What is the sum of *all* of the probabilities, both in the one-die and two-dice cases? (The total of $\frac{1}{2}$ and $\frac{1}{2}$ is 1, as is the sum of $\frac{1}{4}$, $1\frac{1}{2}$ and $\frac{1}{4}$.)

Step 8

'This is ridiculous!' Augustus yelled. 'You keep making wonky dice. You've got to have opposite faces that add up to the same amount.'

Why does Augustus want opposite sides to add to the same amount? (It helps the die to be balanced and so less biased.)

Ask your students which of the dice so far are not 'wonky', according to Augustus.

Looking at the dice in Step 4, there is only one die whose opposite faces all add to the same number (3)—the one where opposite faces have a one and a two.

Incidentally, if we have three faces with the number 1 and three with the number 2, the sum of all the faces is nine. Since there are three pairs of opposite faces on a cube, the sum of opposite faces we are looking for is $\frac{9}{3} = 3$. You can use this method to prove that all three pairs of opposite faces cannot add to the same number if we have two faces with the number 1 and four with the number 2. ($2 \times 1 + 4 \times 2 = 10$ and $\frac{10}{3}$ isn't a whole number.) The same is obviously true for one face with the number 1 and five with the number 2.

Where to from here?

- Can your students give examples of being systematic and using symmetry in their work? Did anyone make a false conjecture? What was the hardest thing they had to do here?

- Suppose that Julius rolled a combination of two of any of the dice he had made so far. What are the possible sums and the probabilities of getting these sums? Are any of the dice 'fair' in the sense that the probabilities of each sum are the same?

- Students select any die they like from above. What sums can the students get from three dice? What are the probabilities of these sums? Extend this to four or five dice and so on. Can they see any patterns?

Level 2: More Roman dice

Problem

Augustus was very frustrated with Julius' dice experiments. 'Not only are most of your dice wonky but they're also of little use. Who ever heard of true Romans gambling with dice that only have 1s or 2s on their faces?'

'Very well', Julius replied, 'suppose that we try dice with 1s, 2s and 3s?' Before Augustus could respond, Julius started cutting up another die. This one had one face with 1, two faces with 2 and three faces with 3.

How many different dice could Julius make? Do any of these have all pairs of opposite faces adding to the same number?

Problem steps

Step 1

We are beginning to move beyond the use of nets, although drawings and concrete materials will still help to produce answers.

After experimenting, students should come up with a conjecture for the number of possible dice. Let them discuss their justifications in their groups before having a class discussion and verification session.

Because of the symmetry of the cube, we can assume that the one is placed on the cube's top face. Then, as shown in Figure 3.6, the bottom face can only be a 2 or a 3.

Figure 3.6: The numbers 2 or 3 must be opposite the face with the number 1

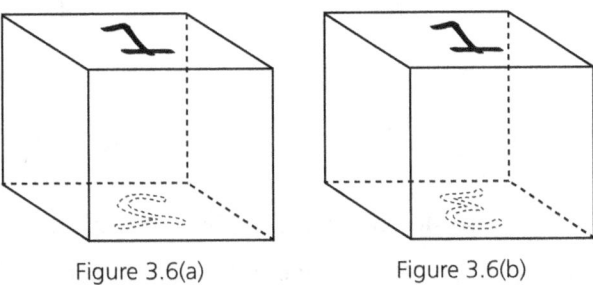

Figure 3.6(a)   Figure 3.6(b)

In Figure 3.6(a) the remaining two has to be on a side face. Since each side face is the same (by rotational symmetry), there is only one die we can make.

In Figure 3.6(b), we have to put two 2s and two 3s on the side faces. By using symmetry we can see that the 2s can only be put opposite each other or next to each other. This gives the two distinct dice of Figure 3.7.

Figure 3.7: The die with the numbers 1 and 3 on opposite faces

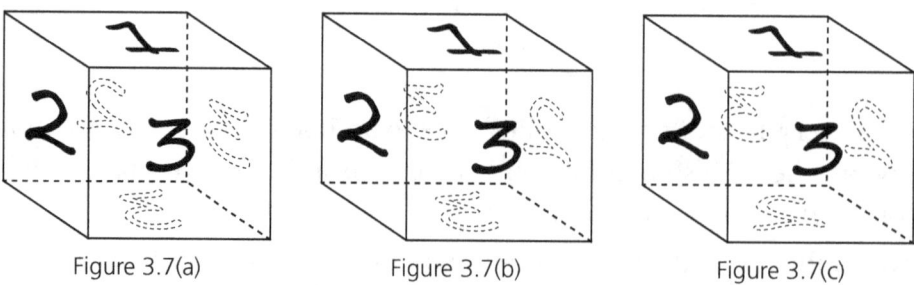

Figure 3.7(a)　　　　　　Figure 3.7(b)　　　　　　Figure 3.7(c)

Therefore there are three dice that have one of number 1, two of number 2 and three of number 3.

Step 2

What is the probability of getting a 1, 2 or 3 with each of Julius' dice? (For a 1: $\frac{1}{6}$; for a 2: $\frac{2}{6}$; for a 3: $\frac{3}{6}$.)

If Julius made two dice with any of these patterns, what sums would he get when he rolled them? What probability would each of these sums be? (For any combination of dice the sums are 2, 3, 4, 5 and 6. The probabilities are for a 2: $\frac{1}{36}$; for a 3: $\frac{4}{36}$; 4: $\frac{10}{36}$; for a 5: $\frac{12}{36}$ and for a 6: $\frac{9}{36}$.)

What is the sum of all of the probabilities, both in the one-die and two-dice cases? (The total of $\frac{1}{6}$, $\frac{2}{6}$ and $\frac{3}{6}$ is 1, as is the sum of $\frac{1}{36}$, $\frac{4}{36}$, $\frac{10}{36}$, $\frac{12}{36}$ and $\frac{9}{36}$.)

Step 3

Unless opposite faces add up to the same number, Augustus thinks they are 'wonky'. How many of the dice Julius made are not 'wonky'?

By looking at each die separately we can see that all the dice are wonky. However, we can also do it numerically because $1 \times 1 + 2 \times 2 + 3 \times 3 = 14$ and 14 is not divisible by 3 (see Level 1, Step 4).

Step 4

Before Augustus could recover from his confusion, Julius started on another type of dice. This time he tried making dice that had two faces with 1, two faces with 2 and two faces with 3; maybe there would be some 'non-wonky' dice among this lot.

How many dice are there that have two faces with 1, two faces with 2 and two faces with 3? Do any of these have all pairs of opposite faces adding to the same number?

While we still expect students to start this problem using concrete materials or drawings, some of your students may now be able to go straight to using a systematic approach. They should then check with other students to see if they all have the same dice or not. When they can't find any more by just searching, they should then make a conjecture and move to a justification.

Working systematically and using symmetry, we can assume that the first number 1 is on the top face of the cube. This allows 1, 2 or 3 to go on the bottom face.

Case I: If we have a number 1 on the top face and a number 1 on the bottom, then the first number 2 can be put on any of the side faces. The next number 2 can either go next to the first 2 or opposite it. The seemingly different ways of putting the two number 2s on adjacent faces are just rotations. Then the number 3s can be placed on whatever faces are left. This means that there are two ways that the number 1s can be opposite each other, as shown in Figure 3.8.

Figure 3.8: The die with two number 1s on opposite faces

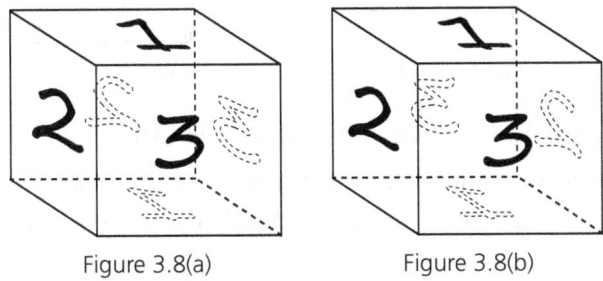

Figure 3.8(a)   Figure 3.8(b)

Case II: Suppose that we have a number 1 on the top and a number 2 on the bottom (see Figure 3.9). Symmetry again allows us to put the next number 1 on any face, so we put it on the right face.

Figure 3.9: A die with number 1 on the top face and number 2 on the bottom

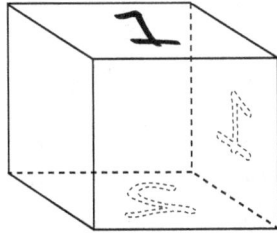

The other two could go on any of the three remaining faces. Are these all different? Can we find a symmetry to show that two of these are the same?

If the number 2 is on the left face then we have two pairs of opposite faces with a 1 and a 2. Adding the two number 3s gives us the die in Figure 3.10.

Figure 3.10: A die with numbers 1 and 2 on two pairs of opposite sides

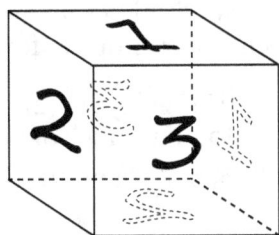

Figure 3.11 shows the second number 2 on the front face (at left) and then the back face (at right). So we have at least two dice at this point. We showed the two cubes with only one pair of opposite faces with numbers 1 and 2 in Figure 3.8. Because there is only one pair of opposite faces with numbers 1 and 2, the only possible symmetry of these dice is a rotation around a vertical axis. But this symmetry also has to fix the opposite pair of the number 1 and 3. So the two dice in Figure 3.11 are different.

Chapter 8: Roman gamblers   121

Figure 3.11: The dice with the number 2 first on the front face and then on the back

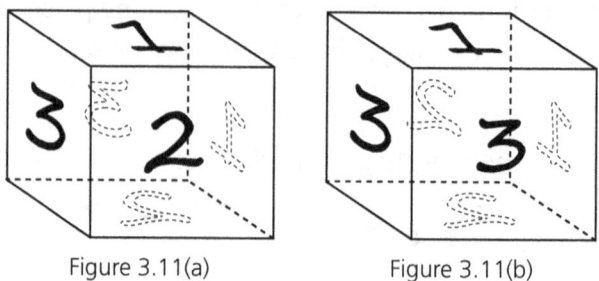

Figure 3.11(a) Figure 3.11(b)

From Figures 3.10 and 3.11 we see that Case II gives three dice.

Case III: There is a number 3 on the bottom face. In this case we can interchange the numbers 2 and 3 from Figures 3.10 and 3.11. We show these in Figure 3.12.

Figure 3.12: The dice with number 2 first on the front face and then on the back

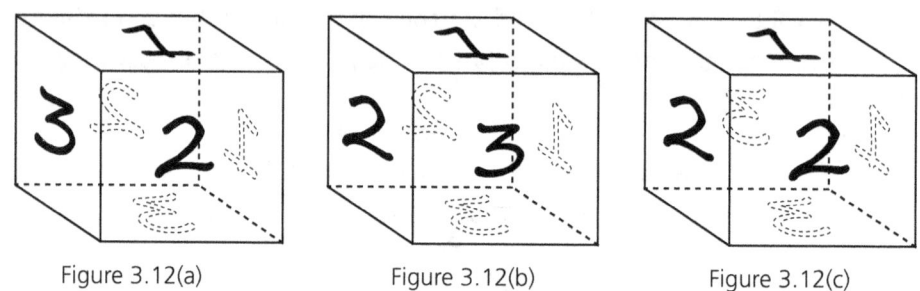

Figure 3.12(a) Figure 3.12(b) Figure 3.12(c)

Figure 3.12(a) is the only die so far that has no pair of opposite faces with numbers 1 and 2, so this must be new. But the other two dice have a pair of faces with 1 and 2 opposite each other, and we already found these in Case II.

Therefore Case III produces only one new die and there are six dice altogether.

Step 5

Get the students to calculate the probabilities of using any one of these dice once, twice or three times.

1. One dice: probabilities of 1, 2 and 3 are $\frac{2}{6}$.
2. Two dice: probabilities of 2: $\frac{4}{36}$; 3: $\frac{8}{36}$; 4: $\frac{12}{36}$; 5: $\frac{8}{36}$; and 6: $\frac{4}{36}$. (By now, students should be able to reduce these fractions to their simplest form.)
3. Three dice: probabilities of 3: $\frac{8}{216}$; 4: $\frac{24}{216}$; 5: $\frac{48}{216}$; 6: $\frac{56}{216}$; 7: $\frac{48}{216}$; 8: $\frac{24}{216}$; and 9: $\frac{8}{216}$. These are a little harder to calculate. The total number of outcomes is $6 \times 6 \times 6 = 216$, but how many different rolls give 3, 4, 5, 6, 7, 8 and 9?

Perhaps the easiest way of looking at this is to assume that you have rolled two dice. With two dice, 2 appears 4 times, 3 appears 8 times, 4 appears 12 times, 5 appears 8 times and 6 appears 4 times. This means that:

- 3 occurs 2 × 4 times = 8 times (from the 1s on a single dice and the 2s on two dice)
- 4 occurs 2 × 8 + 2 × 4 = 24 (from the 1s on a single dice and the 3s on two dice, as well as the 2s on each)
- 5 occurs 2 × 12 + 2 × 8 + 2 × 4 = 48
- 6 occurs 2 × 8 + 2 × 12 + 2 × 8 = 56
- 7 occurs 2 × 4 + 2 × 8 + 2 × 12 = 48
- 8 occurs 2 × 4 + 2 × 8 = 24
- 9 occurs 2 × 4 = 8

Step 6

How many dice are there with all pairs of opposite faces adding to the same total?

Since 2 × 1 + 2 × 2 + 2 × 3 = 12, then that sum ought to be 4. Figure 3.11(a) gives the only possible die.

Where to from here?

- What was the hardest thing to do here? Was the hard part being the systematic part or using symmetry?
- You might like to look at the number of dice that have the numbers 1, 2, 3, 4 and two 5s. What are probabilities for four or more dice? Is there a pattern?

Level 3: How many proper dice?

Problem

How many different ways can Julius put the numbers 1, 2, 3, 4, 5 and 6 on the faces of a cube to make a 'normal' die?

Problem steps

Step 1

Most students will think that the answer to the problem has to be that there is just one way. Others will think that you would not ask the question if the answer was that simple. 'One' is the obvious answer, but this is maths, and maths rarely defaults to the obvious answer. (It's a bit like some of Stephen Fry's questions on *QI*.)

Whatever students say, ask them for their reasons for their answer. Let them go away and try to construct as many different dice as possible that fit the problem's parameters.

Step 2

After students have worked on the problem for a while, let them discuss their findings as a class.

They should have found that there are two possible 'normal' dice, but there may be a variety of answers. If so, encourage the class to find a *systematic* way of approaching the problem before they set off. You may have to give them some idea of what you mean by this, possibly by showing them the start of the argument in the next step.

Step 3

(It would be ideal if some of the students could come up with this argument and show the others. Allow comments or questions as they give the solution.)

We will work this out by looking at where the numbers 1 to 6 could go. It might be a good idea for students to have some blank cubes/templates so that they can follow along with you.

Let's start with the number 1. The 1 can be put on any blank face, but once we have chosen a face for the 1, the 6 must go on the face opposite that, as shown in both the picture and the net of Figure 3.13.

Figure 3.13: The numbers 1 and 6 on opposite faces of the cube

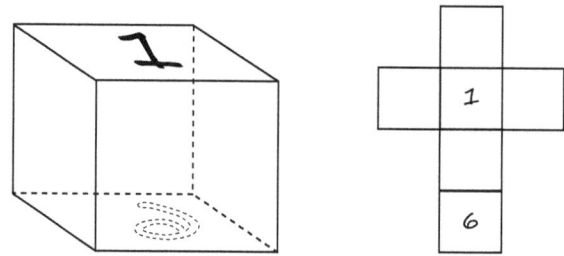

Where can the number 2 go? It has to go on one of the side faces in Figure 3.13. At the moment there is no difference between these faces—to see or show this, rotate the cube in space. So we can put the 2 anywhere we like. Then the 5 goes on the opposite face, as shown in both the picture and the net in Figure 3.14.

Figure 3.14: The numbers 1 and 6, and 2 and 5 on opposite faces of the cube

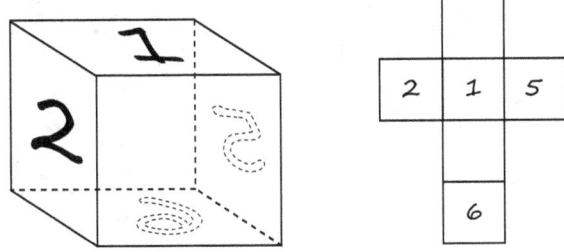

Finally, we have *two* choices for the number 3 (with the 4 then going onto the face opposite the 3). The number 3 is either on the front face of the cube or the back face. We show these choices in Figure 3.15 below.

Figure 3.15: All of the numbers on the cube

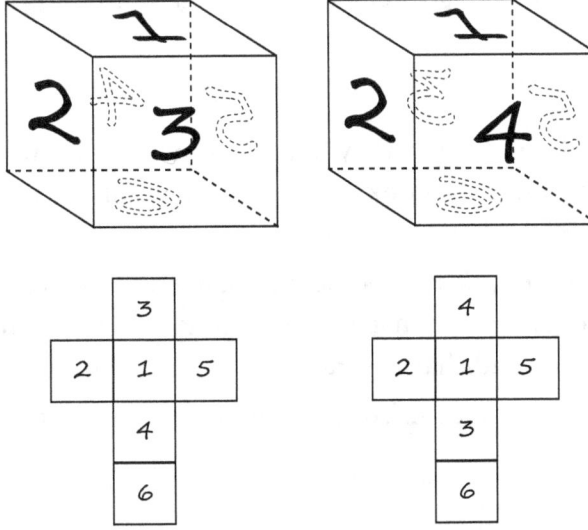

Step 4

The question is: does it really matter? Have we produced one single object or two slightly different ones? Do we have one or two normal dice?

What does the class think? Is there only one normal die or are there two?

Give students a few blank cubes or cube diagrams that they can write on, then send them off to work on this in their groups. (Printable cube and nets diagrams can be downloaded from the series website). They should eventually make the two dice of Figure 3.15 .

Step 5

Get the students to discuss why there are actually two different normal dice, and why the difference is meaningful.

To see that there is a difference between the two options, look at the corner of the cube that is common to the faces showing numbers 1, 2 and 3. We do a bit of rotating to go from Figure 3.15 to Figure 3.16.

Figure 3.16: Looking at the two cubes from the 1, 2, 3 corner

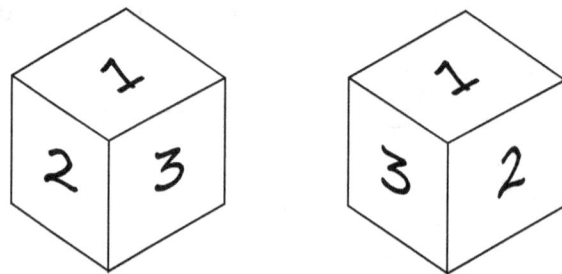

Have we created one die or two dice? We will have only one if we can rotate one of these dice to fit it over the other one—and this can't be done. No matter how hard you try, you can't get the faces to fit onto one another. So there are actually two possible normal dice.

Another way is to look at the dice so that the 1 is right in front of you. Then what can you say about the numbers 2, 3, 4 and 5? On one die the 2, 3, 5 and 4 are in a clockwise direction; on the other they are anti-clockwise. So again, this shows that there are really two different dice.

Step 6

Let the class bring to school the dice they have at home. Can they find both possible dice in this collection? They could also look up the properties of 'normal' dice online to see if two different dice are shown there.

Discuss the concept of *chirality*; in geometry, a figure or solid is chiral (and has chirality) if it can't be mapped to its mirror image solely through rotations and translations. What other kinds of shapes and solids have chirality?

Make some large permanent dice. The best might be shown the next time parents come to school.

Step 7

Are these dice fair? Do you get the same probability of getting any number? How can you show that?

Is rolling two dice a fair event? Do the numbers 2 to 12 come up equally often? Why? Why not? If not, what are each of the probabilities? (Probability of 2 and 12 is $\frac{1}{36}$; of 3 and 11 is $\frac{2}{36}$; of 4 and 10 is $\frac{3}{36}$; of 5 and 9 is $\frac{4}{36}$; of 6 and 8 is $\frac{5}{36}$; of 7 is $\frac{6}{36}$.)

What about three dice? What sums can you get? What are each of the probabilities? We list the probabilities in Table 3.6.

Table 3.6: Outcomes and probabilities of rolling three dice

Number	3	4	5	6	7	8	9	10
Probability	$\frac{1}{216}$	$\frac{3}{216}$	$\frac{6}{216}$	$\frac{10}{216}$	$\frac{15}{216}$	$\frac{21}{216}$	$\frac{25}{216}$	$\frac{27}{216}$

Number	11	12	13	14	15	16	17	18
Probability	$\frac{27}{216}$	$\frac{25}{216}$	$\frac{21}{216}$	$\frac{15}{216}$	$\frac{10}{216}$	$\frac{6}{216}$	$\frac{3}{216}$	$\frac{1}{216}$

Ask students how they think these probabilities can be worked out. One way is to use a table to list every combination that adds up to a specific number—one combination adds to '3', three combinations add to '4' and so on.

Another is to list all the combinations with ones first: (1, 1, 1), (1, 1, 2) etc.; then five more—(1, 2, 1), (1, 3, 1) etc.; then five more—(2, 1, 1), (3, 1, 1), etc.; then five more—(1, 2, 2), (2, 1, 2) etc.; and so on. This does get tedious after a while though.

Where to from here?

- Who in the class had a simple answer for this problem? Can they find a simpler way of getting the answer here?
- How did they use symmetry?
- What games use dice? What games use more than two dice? What games use dice that are not cubes?
- Construct and analyse triangular prism dice. See how many there are and what each of the probabilities is.
- What other dice can your students construct? What probabilities do each of these dice have?
- What do they think was the hardest thing that they did in this Level?

Level 4: How many wonky dice?

Problem

How many dice are there with each of the numbers 1, 2, 3, 4, 5 and 6, where the opposite sides *may* or *may not* add up to 7?

Problem steps

Step 1

The first step here is the same as in Level 3. Discuss what the students think. It is likely that they will make all sorts of guesses. Whatever they say, ask them for their reasons for their answer. Let them go away and try to construct as many different dice as possible that fit the problem's parameters. This may take some time.

Encourage them to report back on the numbers of dice that they found and the reasons for those numbers. Although it is a biggish number, the class could make up all of them and put them on display.

It's vital in this problem for students to work *systematically* in some way, otherwise they will get totally lost and unable to see if all of the dice have been listed. Emphasise the approach that was taken with normal dice and encourage the students to do the problem with some order in mind.

Let them discuss their results. As there are many possibilities, ask them how they made sure they got every single one. How were they systematic? Is there any consensus in the results? Did they use the methods used in all three previous levels?

Step 2

(It might be helpful for students to have a blank die to hand when you go through this. Encourage different students to make suggestions at each stage. Make sure that the rest of the students are happy with what is said.)

As with the normal dice, the number 1 can go anywhere we like to start with. There are then five choices for the number that goes opposite the 1 (see Figure 3.17 below).

Figure 3.17: Five possibilities so far

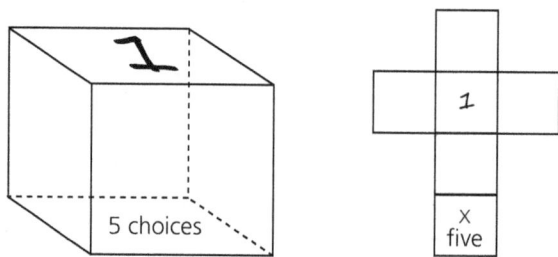

Now you can spin the cube around between finger and thumb and the four faces that have no number on them are indistinguishable. So put the next number, X, on any face. In Figure 3.18 we have X on the left face.

Figure 3.18: All the possibilities

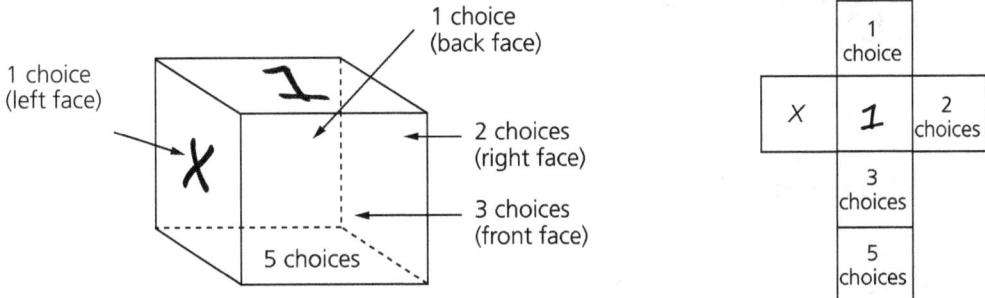

Now we only have three numbers to assign to three faces. If we put the first number on the front face, there are three choices for this number. The remaining faces are different, so we have two choices of numbers for the next face (the right one) and one choice for the final face (at the back).

In total, we have five choices by three choices by two choices by one choice. So we have $5 \times 3 \times 2 \times 1 = 30$ possible dice.

Step 3

Distribute blank cubes and templates among the student groups. Let the class decide how to divide the 30 possibilities among the groups. For example, if you have five groups, they might be assigned the dice with 2, 3, 4, 5 or 6 on the face opposite to the 1.

When the dice are finished, get the class to check that they are all there.

At this point you might like them to produce permanent versions, which could be displayed if the class has an open day.

Step 4

Ask: How does the probability of throwing a 6 change, depending on which of the dice are used?

(It doesn't; no matter how the faces are distributed, they all give the same probabilities.)

Where to from here?

- What was the value of being systematic in this Level?
- It would be worthwhile to produce a class set of 30 dice. Which ones are the two normal dice?
- How many dice can they make using triangular prisms? What about prisms in general?
- There are many other types of dice, not based on a cube, which you can use as the basis for extending this problem. You might also count the number of dominoes that you can have. See what the class comes up with.
- See what your students can find out about cubes online. Once again you might get them to tell you about their most interesting find or their most interesting picture.
- In this activity, what was the hardest thing that they did?
- What do they think that they learnt by doing this activity?

CHAPTER 9:
THE SCHOOL FETE

Initial problem

The school fete is coming up in two months' time and Grade 5C is planning to help by making and selling cakes. Is this worth the effort or would it be easier to do something else? If it would be easier to do something else, why is that the case? If it is worthwhile, what profit is Grade 5C likely to make from their cake stall?

Background information

The overall idea of this activity is to prepare for engaging in a profit-making activity at the annual school fete. This involves both financial mathematics and the collection and representation of data in a statistical way. This makes it a cross-strand activity involving both data representation and interpretation (Statistics and Probability) and money and financial mathematics (Number and Algebra).

The activity could also involve the integrated studies, art/design and economics sections of the school. Opportunities for cooperation arise because: (i) recipes will need to be found, ingredients bought and cooking carried out; (ii) art has to be produced to advertise the cake and sausage sizzle; and (iii) a marketing program will have to be developed.

In Level 1, students investigate the popularity and profitability of certain cakes, and make a Microsoft PowerPoint presentation of their findings to show how valuable selling cakes at the fete might be. Level 2 is a repetition of the theoretical work of Level 1, but focusing this time on a sausage sizzle. Both are suitable for all middle/upper primary students, but younger students could not be expected to work at the same depth as older students.

Level 3 brings the results of Levels 1 and 2 together, and students make a case for either a cake stand or a sausage sizzle at the fete. This case needs to be presented to the fete committee or the principal. Level 3 is best used for students in Level 5 and above, but many Year 4 students would gain from looking into the ideas here.

Finally, in Level 4, students are asked to collect data at the fete to see what their actual sales and profits were and compare them to the predictions that they made in their presentations in Level 3. Level 4 is best for students in Years 6 and 7, though more able students from Year 5 could find it valuable.

Table 3.7: Australian Curriculum content descriptions for the *School fete* activity

Activity level	Problem	Content descriptions
1	The cake stall	*Year 4* Solve problems involving purchases and the calculation of change to the nearest five cents with and without digital technologies (ACMNA080) *Year 5* Pose questions and collect categorical or numerical data by observation or survey (ACMSP118) Construct displays, including column graphs, dot plots and tables, appropriate for data type, with and without the use of digital technologies (ACMSP119) Describe and interpret different data sets in context (ACMSP120) Create simple financial plans (ACMNA106)
2	The sausage sizzle	*Year 4* ACMNA080 (see above) *Year 5* ACMSP118 (see above) ACMSP119 (see above) ACMSP120 (see above) ACMNA106 (see above)
3	Cakes or sausages?	*Year 5* ACMSP118 (see above) ACMSP119 (see above) ACMSP120 (see above) ACMNA106 (see above) *Year 7* Investigate and calculate 'best buys', with and without digital technologies (ACMNA174)

Table 3.7: Australian Curriculum content descriptions for the *School fete* activity

Activity level	Problem	Content descriptions
4	The washing up	*Year 5* ACMSP118 (see p. 131) ACMSP119 (see p. 131) ACMSP120 (see p. 131) ACMNA106 (see p. 131) *Year 7* Calculate mean, median, mode and range for sets of data. Interpret these statistics in the context of data (ACMSP171) Describe and interpret data displays using median, mean and range (ACMSP172) ACMNA174 (p. 131)

Note that at some stage during this activity you will need to talk to your students about any potential allergies. It may make sense to make the ingredients 'nut free'.

Before starting this activity, look at the video 'Hans Rosling and the magic washing machine' (see the series website for a link) to get some ideas for representing data. This may help you to help your students find novel ways to present the data from which they need to make decisions. For example, in Figure 3.20 we represent the profit on a cake by using its icing layer; the 'cakey' part of the figure is the outlay spent to get the cake to the fete for sale.

Big ideas

Create simple financial plans; display data; analyse and interpret the data; communicate the results; design a marketing campaign.

Suggested resources

» Cake ingredients (these will depend on what things you decide to make and sell)
» Sausages, bread rolls, tomato sauce and related barbecue foods

Problem aims

» To create simple financial plans
» To involve students in real applications of theoretical ideas of profitability and data representation

Key concepts

» Use of tables and graphs
» Creating and testing theoretical models

Possible heuristics/strategies

» Draw graphs
» Draw tables
» Collect data

Level 1: The cake stall

Problem

The school fete is coming up in two months' time and Grade 5C is planning to help by making and selling cakes. Is this worth the effort or would it be easier to do something else? If it would be easier to do something else, why is that the case? If it is worthwhile, what profit is Grade 5C likely to make from their cake stall?

Problem steps

Step 1

You may decide that using both Levels 1 and 2 with all of the class is a little repetitive. In that case, you could have one half of the class work on selling cakes (Level 1) and the other half on the sausage sizzle (Level 2). Then competing groups can provide a case in Level 3 for which activity the class should organise.

First, discuss the idea of a cake stall. What are students' favourite cakes? What cakes would be easy to make? How many different cakes would be needed to give potential customers a range of options?

Step 2

Get the students to produce a potential list of cakes for the stall. They then need to collect data on which cakes are liked best by students in the school. But how will they go about this? How will they decide which cakes are most popular? They need to discuss this. What possible ways are there to produce their questionnaire?

Here are some basic survey methods:

Method 1: Students list the possible cakes, then ask which cake each respondent prefers. The data is collected by placing a tick against the cake listed by a respondent. At the end, the number of ticks will give an indication of the most popular cake overall.

Method 2: Respondents list the cakes from 1 to however many, where 1 is most popular and the largest number is least popular. (Students might prefer respondents to give the highest number to the most favoured cake; this might seem a more natural way to do things.)

Method 3: Another way (the Likert scale) is to ask students how much they like each cake. We give an example of how this might look in Figure 3.19. For each cake, the respondents say where on a scale they would place each cake.

Figure 3.19: A Likert scale approach

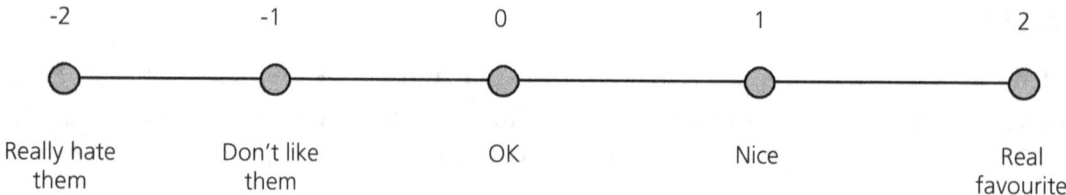

In the discussion regarding collection methods, your students need to think about how they would analyse the results.

Method 1 is easy to analyse; students just count the number of times a given cake is chosen by the respondents. The ones occurring most often are the popular ones that should be considered for the fete.

Method 2 is a little more complicated to analyse, but might give a better indication of the overall feeling of the students questioned. Add up all of the numbers that have been assigned to each cake; the *lowest* combined numbers give the most popular cakes. (You can see why the alternative method might be more natural; in this case, the popular cakes have the *highest* sums.)

Method 3 perhaps gives the best representation of how students feel about the cakes. It again requires all of the scores for each cake to be added up; the cake with the highest score is the most popular. This is slightly more complicated than Method 2 but gives a better idea of respondents' preferences.

Step 3

The question to ask now is how to display the data so that it is possible to make the best decision on what cakes should be on sale. This requires another discussion.

There are many ways to display the data. It may be possible to see the best choices simply by drawing up a table or producing a bar, column or line graph. An Excel spreadsheet can be used to produce these.

Note that a bar graph is *not* the same as a histogram. A bar graph is only used for data where at least one of the variables is *discrete*, or a whole entity. Since cakes are different from each other, they are discrete. The height of the bar is the number of times the given discrete variable occurs. The width of the bar is not relevant to the statistics (of course, the wider it is, the easier it is to see). Bar graphs are also referred to as column graphs or line graphs—although the label 'line graph' is more commonly used for a graph made by joining together points of data.

Histograms are used when the variables are *continuous*. This means that they have values that vary over a range—for instance, they can be any decimal number from 1 to 2. Continuous variables can normally be expressed with a unit of measurement. Height of students is a typical example of a continuous variable where histograms are appropriate. The same is true for elapsed time, weight, amount of energy etc.

Step 4

The students should look at the data and decide which cakes they will make—perhaps the five most popular types. But how many of each cake should be made?

This will depend on the popularity of the cake, how many people are likely to attend the fete, how big the stall is going to be and so on.

Let the class discuss this in groups and then make a decision as a whole class. They should bear in mind that they might be able to charge more for some of the less popular cakes. So even if they sold fewer cakes, the students might make more money on these than on other cakes.

Step 5

The next thing to find out is the potential profit on each cake. How could the class go about this task?

This seems like a matter of adding together the prices of all of the ingredients. But some ingredients come in packages that will make more than one cake, so how many of each cake will the students bake? How many will they hope to sell? Why? On the other hand, they may decide to buy packets of ingredients that make a particular cake. This is likely to increase the costs, though. Why?

Students then need to decide how much to charge per cake. They might get some idea of the price of corresponding items in the supermarket or school canteen/tuckshop. They should also realise that at the school fete, people may be willing to pay a little more as they are trying to support the school. Students might consult teachers or parents about what they would be prepared to pay.

Step 6

We have assumed so far that the baking might be done at the school as part of the integrated studies program. How will it change your pricing structure if some parents donate the ingredients?

Step 7

Now your students can prepare their case for how much money they expect to get for each type of cake. How can they display this data?

One way to display the data is to prepare sketches, like the one in Figure 3.20, to put on a pictograph. Here the height of the body of the 'cake' represents the cost of producing it, while the 'icing' is the profit expected.

Figure 3.20: The expected profit per cake

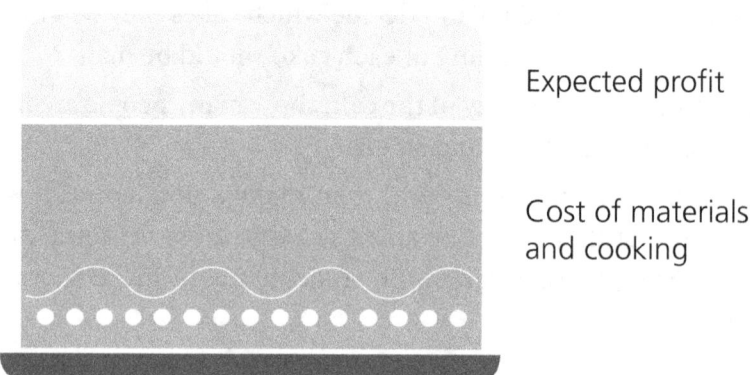

All of the variables involved need to be considered, as these will affect pricing and profitability. Different groups can now present their projected profits per cake to the class. This should be presented formally and needs to include information about:

- initial investment
- list of expenses
- projected revenue
- projected profit
- margin for error. (For instance, what if there are fewer sales than anticipated?)

Step 8

From this data, is it worth running a cake stall?

That will depend on the amount of profit expected overall, so it needs to be calculated. How much money is likely to be made by the cake stall? Is it worth all the effort involved?

Get the student groups to prepare PowerPoint presentations that put forward the cases for and against running a cake stall on the day of the fete.

Where to from here?

- Could the students increase their profit by getting different people or businesses to sponsor cakes? How will this increase their potential profit? Should they decrease the price of the cakes? Will they make sure that their sponsors are listed on the stall?
- What will they do with any cakes that are left over after the stall?
- What other things can your students think of to produce and sell?
- Can they think of ways to improve the approach that they have taken at this level?

Level 2: The sausage sizzle

Problem

The school fete is coming up in two months' time and Grade 5C is planning to help make some money for the school by cooking (on the day) and selling sausages, hamburgers and so on. Is this worth the effort or would it be easier to do something else? If it would be easier to do something else, why is that the case? If it is worthwhile, what profit is Grade 5C likely to make from their sausage sizzle?

Problem steps

Similar steps and methods apply here as were used in Level 1, 'The cake stall'; follow the same steps that were outlined there. The differences come from noting what things apply to the sausage sizzle that don't apply to the cake stall.

Here are some questions to pose to the students:

- What kind of sausages will you sell? Will you sell just one kind, or will you offer different types? Should you charge more for different types of sausage?
- Will you sell hamburgers as well as sausages? How much will you charge for them?
- You need to have bread to serve the sausages in. Will you offer just one kind, or have the option of white, wholemeal etc.? What about buns for the hamburgers?
- Will you have onions to put on the sausages? What about extras for the hamburgers—are you offering just a burger in a bun, or will you have salads, cheese etc.? Will they be free or cost extra? What about sauces?
- What prices are typically charged by other sausage sizzles in your community?
- The sausage sizzle will need sausages/hamburgers to be bought, rather than made. How much will they cost? Can you get them cheaper from a different supplier?
- An adult will have to do the cooking. Can you get enough volunteers to do this? How will you arrange for volunteers?
- What other issues need to be taken into consideration?

Where to from here?

- What other ways can your students think of to raise money at the fete?
- How could the class improve their collection and analysis of the data?

Level 3: Cakes or sausages?

Problem

Should the class run a sausage sizzle or a cake stall—or both—on the day of the fete?

How will you make this decision?

Problem steps

Step 1

Divide student groups into pro-cake and pro-sausage lobbies. Each lobby should revise the PowerPoint presentations that were prepared in Levels 1 and 2 as needed. They should think about the objections they might make to the other lobby's arguments, as well as the objections that the other lobby might use to their own case. Their objections should be backed as much as possible by graphic displays and logical arguments.

Step 2

Each lobby should present their argument and PowerPoints to the class. This could be done in a Parliament-like situation, with the 'cakes' to one side and the 'sausages' to the other. Try to involve as many students as possible in 'active' roles (Speaker, Clerks etc.).

At the end of the presentations, hold a vote to decide which way the class should proceed (or whether the class should support both or neither activity).

Step 3

Assuming that the class votes to proceed on one or more of the ventures, they now need to design a marketing campaign.

The campaign should be divided into two parts. First, the class needs a campaign for the period leading up to the fete. This is probably going to be aimed at both the students in the school and their parents. The second part is for the day of the fete, making sure visitors know about the cake stall and/or the sausage sizzle.

Step 4

How could the PowerPoint presentations be improved? This should first involve a whole-class discussion, followed by groups working on particular issues.

Prepare a final report to put to the fete organisers. This should involve a PowerPoint presentation with at least three different types of data display, plus the plan for a marketing campaign to attract people to the stall.

It might be useful to include how the class will dispose of rubbish, how and when school equipment will be cleaned and returned to its appropriate place etc.

Where to from here?

- You might do a trial run on both the cakes and the sausages before the fete actually takes place. Even if your class is not involved in the fete, you could have a mock trial run one lunch time for the other students and staff in the school.

Level 4: The washing up

Problem

How will the class know if they have made a profit or not? How does this compare with the predicted profit presented to the fete organisers?

Problem steps

Step 1

What steps need to be taken on the day of the fete to ensure that the class knows if there has been a profit or loss?

The simplest way to do this is just to make sure that students are accurate in receiving money and giving change. However, if we want to compare the actual outcome against that predicted, then someone needs to keep track of how many of each item is sold. This will have advantages for another year when more accurate ordering of stock for the stalls can be made. This can be done by keeping track of the sales during the fete and then checking it against the difference between stock made or purchased and stock left at the end of the day.

Step 2

Ask the class to show how the actual profit or loss might be compared with the predicted profit from Level 3.

This data can be presented in at least two ways. One way is as a bar graph on which two bars are given for each type of item. One bar shows the predicted profit and the other the actual profit (which may be negative—shown below the line).

Part of such a graph is shown in Figure 3.21.

Figure 3.21: Showing predicted profit against actual profit

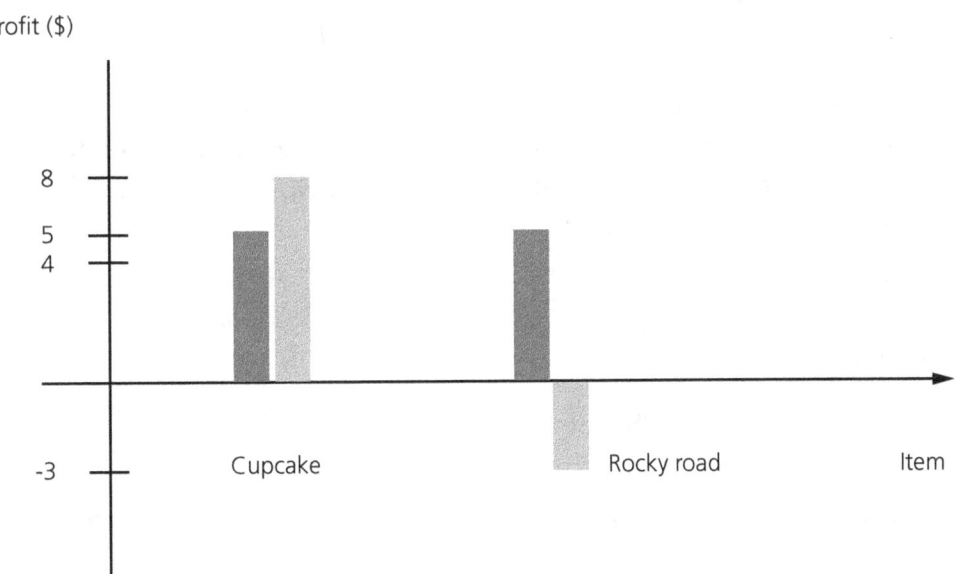

Chapter 9: The school fete

Another way to show this is as a balance sheet with any losses marked in red. This could be created using Excel or a similar spreadsheet program.

Step 3

Can your class think of any other way to present this comparative data?

This could be done using tables by putting the bars from the actual sales on top of the bars for the predicted sales, by superimposing a line graph for the predicted profits on top of a line graph for actual profits, and so on. Let them explore possibilities using the various chart generation options in Word and Excel.

Another view on this can be seen by using *mean profits*. The mean profit (as either profit per item sold or profit per item made) of each item can be found along with the mean profit overall of the items. By comparing these means, students can determine which item made the best profit per item.

After groups have found a way they like, they should make up the data presentation and show it to the rest of the class.

Which presentation does the class like best? Which is the easiest to understand?

Step 4

It's unlikely that all of the predicted results for all the items will be *exactly* the same as the actual results, but they may be close in some instances. How close will the class accept as being *essentially* the same? Perhaps an error margin of 10%? What do they think? Why did they choose that level of tolerance?

How many items were within the tolerance level that the class sets?

Step 5

How many items were outside the tolerance level? Can the class explain why those items were different?

For example, was it a hot day and nobody wanted sausages? Did it rain, so not very many people turned up to the fete?

Could any of your research before the fete have drawn incorrect conclusions?

Step 6

Get groups to prepare a report for the 'shareholders' (the school fete committee). This should show all the comparative data, a profit and loss account, and explanations as to why the actual profit and predicted profit were not the same. The report should go into the reasons behind the tolerance limits set for variations.

Present this report to the shareholders.

Where to from here?

- What was the most interesting part of the *School fete* activity?
- What was the hardest part?
- What did you learn as a result of this activity that you think might be useful to you after you have left school?
- What other projects would you feel capable of taking on?

CREATIVE ACTIVITIES IN **MATHEMATICS**

Problem-based learning is a powerful alternative to drill-and-practice or skills-based learning, especially within mathematics.

The *Creative Activities in Mathematics* series provides a wealth of investigations and open-ended active learning activities, designed to engage students with mathematics and develop their problem-solving, collaboration and mathematical skills.

The three titles in the series provide a variety of class activities suitable for students from lower primary to middle secondary, along with teaching notes and staged lesson plans. Each activity is a whole-class investigation with open-ended answers that takes a particular scenario and develops it over multiple levels. This enables it to be used both at different year levels and with students of differing ability in the same class. All activities are firmly grounded in the Australian Curriculum: Mathematics.

Links to extra information, activities and student worksheets are available and easy to access online.

About the authors

Derek Holton is a mathematician and an Honorary Professor at the Melbourne Graduate School of Education.

Cath Pearn is a Senior Research Fellow in the ACER Institute and a lecturer in Mathematics Education at the University of Melbourne.

Duncan Symons is a Lecturer of Science and Mathematics Education at the University of Melbourne.

Charles Lovitt has directed several Australian national and state mathematics projects and is now a consultant and workshop presenter.

Amba Press | www.ambapress.com.au | hello@ambapress.com.au

www.ingramcontent.com/pod-product-compliance
Lightning Source LLC
Chambersburg PA
CBHW081102070526
44584CB00021B/3177